SLEEPS WITH
WOLVES

Also by Neil Bockoven

Moctu and the Mammoth People
The People Eaters
When We Met Neanderthals

SLEEPS WITH WOLVES

Neil Bockoven

RARE BIRD
LOS ANGELES, CALIF.

THIS IS A GENUINE RARE BIRD BOOK

Rare Bird Books
6044 North Figueroa Street
Los Angeles, California 90042
rarebirdbooks.com

FIRST HARDCOVER EDITION

ISBN-13: 9781644285831(hardcover edition)
ISBN-13: 9781644286081 (e-book edition)

Distributed worldwide by Simon & Schuster

For more information, address:
Rare Bird Books Subsidiary Rights Department
6044 North Figueroa Street
Los Angeles, California 90042

Set in Garamond
Printed in the United States

10 9 8 7 6 5 4 3 2 1

Library of Congress Cataloging-in-Publication Data available upon request

To Scott Gorham
A wonderful, valiant friend

Acknowledgements

IN ADDITION TO THE people mentioned in my earlier acknowledgments, whose help and support have continued, I'd like to give further thanks to Jeff Eckert and Steve Vealey, who've offered suggestions and provided access to new research that has surfaced. My publisher, Tyson Cornell, has been encouraging, and he and his associates have suggested changes that made the book better.

It was an honor that Scott Gorham took time to read the manuscript and offer comments and encouragement, even though he was gravely ill. Most of all, my wife Denise has been a huge help. She's not only been supportive, but she's pretended not to notice my many piles of papers and reference books.

Finally, we should all give thanks to the many researchers making astounding breakthroughs as they apply ancient DNA analysis and other incredible new technologies to their studies of paleoanthropology.

Author's Note

Sleeps With Wolves is a stand-alone book that follows the saga of the first two novels, *Moctu and the Mammoth People* and *The People Eaters*. I've skipped ahead a dozen years, because I really wanted to write about the grown-up hybrid, Elka, and well, I'm not getting any younger.

I've always admired authors like Michael and Kathleen Gear that can pump out great novels quickly (they recently finished their 80th), but I don't seem to have that ability. I remember a notable quote from Louis L'Amour, who wrote nearly a hundred western novels. He claimed that he could write a novel sitting in the middle of Sunset Avenue with his typewriter on his knees. "Temperamental I am not," he said.

I strive to be like that, and I don't think I'm temperamental, but I just don't write very fast. I also like the words of another famous author, Isaac Asimov, who, when asked what he'd do if he were told he had only six minutes to live, said, "I'd type a little faster."

I'm actually thankful that I'm not like that. There are too many other things in life to savor and enjoy, too many wonderful distractions. I have more in common with this last quote, which may be the best acknowledgement ever:

> *To my wife Marganit*
> *and my children Ella Rose and Daniel Adam*
> *without whom this book would have*
> *been completed two years earlier*

Although I'm slow at it, I enjoy writing—all the exciting action and adventure, sex and violence, drama and love stories. But my real passion is research about paleo humans. New technologies are just giving us so many new and exciting breakthroughs like these:

1. The first complete ancient genome was sequenced in 2010. Five were completed that year, including the breathtaking discoveries that, yes, we HAD interbred with Neanderthals, and wow—Denisovans, a species we knew nothing about previously, had also interbred with us. Svante Paabo led the teams that made those two discoveries, for which he was awarded the Nobel Prize. Since 2010, the rate of data acquisition and discovery has gone exponential. There were thirty-eight genomes sequenced in 2014, 249 in 2015, and by 2017, David Reich's lab alone was pumping out thousands per year.[1] Discoveries are coming so fast that researchers are having trouble keeping up on publishing all the important findings.
2. Most bone fragments found at ancient human sites aren't analyzed because they can't be identified as human. Tom Higham and Samantha Brown tried a new technique—mass spectrometry—to scan small bone fragments for collagen which can then be screened for slight differences to tell if they are hominin remains.[2] They scanned 1,500 bone fragments from Denisova Cave in Siberia (where Denisovan remains were first identified) and determined that all but one were the remains of cows, deer, and other non-hominin animals. One tiny fragment, weighing less than 6/100ths of an ounce, was human, and gene sequencing showed it to be a FIRST-GENERATION HYBRID FEMALE (hmm—sounds familiar), with a Denisovan father and a Neanderthal mother—wow!
3. In the 1940s, huge geoglyphs were discovered in the high Nazca desert of Peru. The geoglyphs were typically several hundred feet across and depicted animals, humans and other figures. Through the early 2000s, one or two would be discovered each year, mostly by people in aircraft. By analyzing high-resolution imagery, researchers increased the rate of discovery to nearly twenty per year between 2004 and 2020, bringing the total number of known geoglyphs to about 430. More recently, scientists trained an artificial intelligence (AI) system to recognize faint traces of geoglyphs, and they identified and ground checked 303 previously unknown geoglyphs in a single field season.[3–4] Hundreds of other possible geoglyphs were seen but have yet to be verified.

These and dozens of other amazing technologies are now being employed, and the new findings are blowing the doors off previously set-in-stone theories about paleoanthropology.

Several new findings that apply to our story of 45,000 years ago are that bows and arrows were used much earlier than previously believed, [5–14, 72] that wolves were probably domesticated earlier,[15–19] and Paleolithic people almost certainly incorporated constellations into their artwork and likely used them for directions, telling time, and predicting seasonal changes.[20–22]

Paleo people likely had vast knowledge of plants, mushrooms and molds. They knew which were good for eating, which could be used for medicines or hallucinogens, and which made potent poisons[23-32]. The Annotated References at the back will attest that, although some cultures were peaceful, hunter-gatherers were generally prone to fighting with their neighbors.[33–37, 71] Likewise, although some groups were environmentally conscious, there were enough bad actors that the environment often lost out.[38–42]

All the new genetic data pouring in has scientists often changing their opinions. When *The People Eaters* was published, researchers tended to dismiss the notion of red and blonde hair and blue eyes in Neanderthals. Now, the evidence is pointing us back that way.[68, 69, 70]

Probably the one overriding theme that stands out from most of the new data pouring in is that these paleo peoples were more creative and advanced than we used to think. I'd encourage you to peruse the references before starting to read this story. Most key parts are highlighted so it'll just take a few minutes. The research underscores how most of what you'll be reading in this novel was, in some form or fashion, happening way back in the Paleolithic.

I hope you enjoy the story.

Main Characters and Places

- **Moctu**—Leader of the Nerean people (*Homo sapiens*)
- **Nerea**—Name of Moctu's tribe (means *mine* or *my people* in Basque)
- **Etseh**—Large rock overhang that the Nereans use for shelter (means *home* if spelled etxea)
- **Nuri**—Talented, dark-haired woman of the Nerea and Moctu's mate
- **Krog**—Tribe of Pale Ones (Neanderthals)
- **Shiv**—Tribe of violent, cannibalistic Pale Ones (Neanderthals)
- **Elka**—Hybrid daughter of Moctu and a Krog woman (elkartu means *unite* in Basque)
- **Uhda**—Shelter of the Krog
- **Lion People**—Large tribe similar to the Nereans (*Homo sapiens*)

PART ONE

The Sea People Come

Chapter One—The Storyteller

"WHY ARE YOU HERE?" the silver-haired woman croaked in a raspy voice.

"To learn," the three boys and one girl said loudly in unison.

"What will you learn?"

"The story of us all," the adolescents said, once again in unison.

The old woman smiled and nodded. The young ones had been coached well.

"Then I'll tell you the story," she agreed, slurring the last word through a gap in her front teeth. The wrinkles on her leathery skin could not totally obscure a flickering beauty, and it was clear to anyone perceptive that she'd once been lovely. But a long, hard life had taken its toll, and the woman's hands shook as she touched the string of engraved bone fragments that hung around her neck. She peered at the largest fragment for a long moment, then her hand moved to the next bone down, and she studied it.

"What…what are you looking at?" the oldest boy said haltingly, a little intimidated by the woman. He sat shoulder-to-shoulder with his twin, the only girl in the cramped hearth. The two had been born twelve summers before with strikingly light-colored hair which they still had.

"It's a long story, and these beads, these tokens keep me from forgetting parts of it or getting them out of order. Now listen and remember, because one day you'll need to tell your children this story. We're looking way back through time, and it's a story of pain and bravery, of skill and endurance…and love and luck." After a pause, she added, "And the blessings of the Spirits." The old woman looked each one of the young people in the room directly in the eyes, then nodded, pleased with their rapt attention.

"It's back in a time when the land had many different peoples and different tribes. Some were peaceful and some…" The old woman shrugged. "Weren't. Some came from the mountains and some from the sea. And the story today starts with a murder of a boy about your age and a girl, not much older, who went out to avenge his death. And the story goes like this."

It was dark, and the hair on the back of Elka's neck rose even before Misha, her wolf, started a low growl.

They're right here.

Too close.

Elka heard whispers—controlled, deliberate—and the soft scrape of movement. Whoever they were, they'd heard her or Misha, and they were readying themselves.

She touched her thumb lightly to Misha's head, and the wolf went still. Elka studied her surroundings and listened to what her Krog grandmother Rah had called the song of the night. The forest to the north had gone quiet—birds, insects, all of it, like a breath being held.

Then she smelled it. Smoke.

She'd nearly walked straight into their camp.

Now that she knew where to look, Elka could make out the faint glow of a tiny campfire through the dense woods.

A noise—closer now—indicated the killers hadn't just taken defensive positions. They were coming for her. She'd been hunting them, and now, they were hunting her. Elka froze and held her breath for fear of making a sound. They were still some ways off. Maybe they'd stop as they got farther from the light of their fire.

A frigid autumn breeze gusted from the northwest, and she got a stronger whiff of the smoke. Elka realized she was shivering. Not the subdued, early type of shivers, but handshaking, violent ones. Was it the cold wind or nerves? She shook her head and wrapped her only fur—her sleeping fur—tighter around her shoulders.

Another hard freeze coming tonight. She grimaced and shook her head again. Right now, I just need to get away. If I can, that is...then I'll figure how I can get ahead of them. It needs to be me ambushing them, not the other way around. She began silently edging backward.

There'd been two of them. They'd come three nights earlier wanting to kill Elka and her family—especially her father, Moctu. They'd killed Ender, the guard, and they'd almost succeeded in killing her father. He still might die, but it was *almost* for now. Mother Nuri was tending him and Elka's youngest brother, Meko, who might die too. Elka scowled and clenched her teeth at the thought. She closed her eyes trying to clear the image of his clubbed, beaten body, so she could focus on the immediate sounds and smells.

After the attack, the marauders had most of a half night's head start, and it had been foolish of Elka to try to chase them in the dark—little trail to follow and too much chance for an ambush. But the light from a half-moon and a little help from a torch showed that they'd retreated northward, and she didn't want them to get too far away. Elka couldn't continue using the torch. That would be like yelling, "Here I am," the whole way as she trailed them. But the half Pale One–half Nerean girl knew the area well—every hill and creek bed, every boulder and giant tree—and Misha's sense of smell was nearly as good as being able to track in daylight.

Most of the Nerean men and a few women were away on a hunting trip that was to take at least half a moon, and the others were dead, wounded or needed to guard Etseh from further attacks. So, an unwavering and seething Elka decided she and Misha would chase them down, and she'd provisioned her pack even as Mother Nuri and Grandmother Alta tried to talk her out of the dangerous pursuit.

Northward from where she was now, the area was heavily wooded with rocky outcrops, but the forested highlands would soon give way to low, rolling hills and lowlands with tall grasses and shrubs.

They're almost certainly warriors of the Lion People, Elka thought. Haven't been hiding their trail. One has a big hole in his right moccasin on the ball of his foot, right behind his big toe. They're large men, bound to be well trained, good with atlatls. I'm no match for them unless I can set up an ambush. Without realizing it, she'd backed up more than ten paces, all the while gently tugging on Misha's fur to get her to do likewise.

Nereans and the Lion People had been at war intermittently for her whole life of fourteen summers. In fact, many believed that Elka was the cause of the war, and some Nereans shunned her for it. Devu, the Lion People shaman, had prompted the war, saying the half-Pale One girl was repulsive and was sickening to the Spirits, and that she was the cause of all the disease and poor hunting. He foresaw the misfortune worsening as long as she lived. Both she and her protector, Moctu, had to die. Although having someone kill Elka would ease the problem, Devu had instructed that she be taken alive, if possible, so that he could sacrifice her to the Spirits.

Now, as Mother Nuri had taught her, Elka moved like the Nightwalkers—creatures that prowl soundlessly through the forest in utter darkness. She retreated with the breeze, moving downslope, and she was finally far enough to the southeast that her heart slowed to a normal rhythm. Elka closed her eyes as she again analyzed the night sounds.

"There are secrets in the whispering wind if you only listen," Rah had told her many times.

For now, there were no worrisome sounds or smells, and as Elka calmed, her mind drifted, and she felt the sadness coming. It was a debilitating gloominess, and it was happening more often these days. Her head sagged, and her lower lip protruded. She rubbed her forehead hard and squeezed her temples before pushing a thick strand of light brown hair behind her ear.

She shook her head violently trying to avoid the looming despair.

There's just too many… Too many have been killed or wounded on both sides, so something needs to change. I'm just…so tired.

The glow of a high tree shimmering with moonlight brought Elka out of her melancholy contemplation.

Perhaps a good omen from the Spirits.

It was time to move, and she knew just the route to get ahead of the killers.

Chapter Two

"We shouldn't have let her go," Nuri said, the trail of a tear still visible on her cheek. "One young girl chasing two murdering men. We can't lose another…" Her voice cracked as she gazed at the covered body of her son of eleven summers, Meko.

"You couldn't have stopped her," Alta said, wiping a cheek with the back of her hand. "We both tried. But Elka's strong and smart, and she has Misha," she continued with more confidence than she felt.

"I can't believe…not one man stepped up to go with her."

"Well, so many are gone…or dead." Alta's gaze moved to her son's body on the ground near Meko's corpse. "Or wounded."

Nuri nodded and scowled. "Once we get Moctu healthy and able to fend for himself, I'll be out there, and I swear…" She shook her head, and all Alta heard was an unintelligible snarl.

Chapter Three

A COLD MIST HUNG in the morning air, and protected pockets of the forest were still dew covered. Elka raised her head higher, over a rounded boulder, peering downslope through the thin foliage of a hornbeam bush that clung to a crack in the rock. She could see well enough through the bush, but the attackers were less likely to see her. Well-trained and quiet, Misha stayed out of sight near Elka's side. The young wolf's two brothers, Peet and Loebo, had never caught on to that. Although all three wolves looked similar, with dark ears and the white tuft of hair on their forehead, they were very different natured. The males weren't as trainable and obedient—or loving—and more often than not, they'd bark or growl at game during hunting and charge ahead. Two winters ago, as spring neared, both males took off, and Elka hadn't seen either since.

All her wolves except Misha had run off at some point, but only Neska, Misha's mother, had returned. And she had returned pregnant. Elka expected Misha to run off sometime in the future when she went into heat. She just hoped the wolf would come back like Neska had.

As Elka anxiously studied her back trail, she chewed on a ragged thumbnail expecting the marauders at any moment. She knew the land far better than they did. Elka felt a bond, a kinship with each of the hills and creeks, and she remembered the position of every large tree and boulder that could hide a person or bend a trail. And using her skills as a Nightwalker—nearly as sharp as her mother's—she'd gotten ahead of the men she hunted.

Now she waited.

It had taken a day to circle around them, and Elka had monitored their progress northward for two more days as she searched for the best place to ambush them. There were problems with every potential site, but it had to be soon. They were nearing the last of the hills before the lowlands, and Elka wanted to use the elevation to her benefit. Her breath caught as she saw a stirring in the brush near the tree line just across the ravine. It was near a trail that she'd used at first light.

"There they are," she murmured, and even though she was pleased that all was going to plan, a chill went through her. She unconsciously felt the smoothness of her wolf-tooth totem.

Now what? They're coming this way. They must suspect that I'm close. But this is the best chance that I'm going to get—it has to be now.

The men were still behind a brush line, and she couldn't see them well—just their movement. Elka bit her lip and shook her head at the thought of what she was up against.

Two grown men were more than a match for a young woman, even one like her. She was very strong, with broad shoulders and a well-muscled body. Although short compared to other Nereans her age, she could out-wrestle any of them, including most of the young men. She could climb trees faster and throw a spear with an atlatl farther. That was part of the problem—another reason she was alone and didn't yet have a mate, even though her menses had come almost a year ago. Some men and boys, probably most, didn't want a girl who was stronger than they were.

Can't think about that now. Focus.

Elka's mind cleared, because now it was time—time for retribution. On the trail the killers were taking, they were nearing a small clearing one good atlatl throw away. From there, the trail began to move westward, away from her position. It has to be there, she thought, clenching her teeth. Elka touched Misha's head again, warning her to silence. She fitted one of the long darts onto her atlatl and readied two more for quick throws. She checked the area around her for limbs and brush that could be impediments to a good throw.

They were big men, and they paused before moving fully into the open. Elka couldn't make out the features of their faces, as they were painted with splotches of black, white and red. She grimaced at the realization that the war paint meant they were expecting to encounter their pursuers. The lead man was tall and broad-shouldered, and he wore a reddish-stained tunic that had brightly colored, red and white beads forming two crisscross patterns on the front. The other man followed closely behind. He was shorter and stockier and had long, dark hair tied in the back. Elka was surprised that neither man wore the distinctive lion-claw and white shell necklaces of Lion people warriors. The men cautiously studied their surroundings, and the leader's gaze lingered on the boulder and brush where she hid. But after a moment, they moved into the open.

Elka rose and threw.

Chapter Four

"You...you're alright?" Moctu murmured weakly. There was relief in his eyes.

Nuri nodded with tears in her eyes. "Mother Alta," Nuri called out loudly, knowing Alta would be relieved to see her son out of a coma. "Mother Alta—he's awake!"

"The children—where...where are Meko, Koby and Elka?" Moctu tried to prop himself up but fell back onto the furs.

Nuri gently patted his unhurt shoulder, "You remember that Koby went on the hunting trip as a pack boy," she said reassuringly. She opened her mouth to say more but then thought better about it.

Alta rushed in and knelt by Moctu, taking his hand from his undamaged right side. "Oh—it's so good to have you awake again," she whispered, and as she closed her eyes, a tear streaked down her cheek.

Moctu smiled thinly, and his eyes closed. "How...?" he started, but after a deep breath, he was asleep again.

Chapter Five

ELKA HAD ALREADY FITTED a new dart onto her atlatl as her first took the shorter man in the right side of his chest, spinning him around. A thin smile of satisfaction and retribution—as well as relief—marked her face.

I'm going to win this, she thought fleetingly.

Both men cried out, and the taller man immediately charged toward her position, drawing a cudgel from his belt with his right hand and holding his spear in his left. He shrieked malevolently as he zig-zagged up the hill making himself harder to hit with a shaft. Elka remembered all that Moctu had taught her—take a breath, focus and follow through. The warrior was halfway to her when Elka threw, and she knew at once it would hit him.

The big man saw her throw, slowed, and watched the dart speed toward him. It was almost to him when he moved his upright spear sideways and knocked the dart to the right, where it careened out of harm's way. His teeth gleamed as he smiled and continued racing up the hill.

Although Elka knew that atlatl darts could be fended, she was momentarily stunned by the development. Half a breath ago, she was sure that she'd won the battle. Now the assailant was almost to her. Recovering, she touched Misha's hind quarters, and the wolf charged from behind the boulder down toward the warrior.

The man's eyes widened as a snarling wolf that weighed nearly as much as he did, came out of nowhere and barreled downward toward him. But again, the well-trained warrior maintained his composure, braced himself and swung his cudgel.

Misha saw the cudgel swing and nimbly ducked below it absorbing only a glancing blow to her thickly furred back. The big wolf hit the man, and the impact sounded as loud as two rutting bucks charging at full speed. The man fell sideways and partly down the hill into an area of low brush and small boulders. Elka fitted a new shaft to her atlatl, but she had no shot. She watched, worried but riveted, as the two bodies rolled around on the rocky ground. She could hear Misha snarling viciously, and through the dust and brush she saw the wolf's head shaking and wrenching savagely, her teeth clamped on the forearm of the warrior's spear hand. The man cried out in pain and alarm, releasing his spear to grab the wolf's neck fur. In moments he, too, was snarling and cursing loudly.

The two figures rolled again, and Misha was on top, violently ripping and jerking at the warrior's left arm. With his right, the man swung his club again, hitting the wolf a hard blow to the ribs, and Misha yelped, almost releasing the arm. Elka whistled for Misha to break away just as the man pivoted and swung the cudgel again. Elka gasped as the cudgel took the wolf in the left eye, crushing it. Misha fell to the side, limp and lifeless.

Stunned to her core, Elka watched dully as the man pushed the dead wolf—her beloved Misha—roughly to the side and retrieved his spear. He was to his knees before Elka came out of her fog and launched her dart. It was a good throw, but even on his knees, the warrior was able to fend the dart away from him. He sneered as he stood and began moving toward her.

Chapter Six

"Where's Elka? Where is Meko?" Propped on an elbow, Moctu was shouting deliriously as Nuri ran into the hearth.

"Moctu, Moctu," Nuri said soothingly while kneeling down, "You need to eat and rest. Let's get your strength back."

"You told me..." Moctu stopped and shook his head, trying to clear it. "You said Koby was...was safe...he's with the hunters." Nuri nodded, and Moctu closed his eyes. His eyes reopened part way, and he said, "How about Elka...and Meko?"

Nuri had dreaded this moment, wanting Moctu to regain his strength before hearing the awful news. "Let's talk about that later after you've rested a little more and gotten your strength back."

"So...so it's bad," Moctu's eyes searched hers. "Are they...are they both...?"

"No," Nuri said quickly. "Elka's..." She looked up briefly searching for the right words, then back to him. "Elka wasn't hurt in the attack."

"But...but Meko..." Moctu closed his eyes again and took a deep breath trying to steel himself.

Nuri had thought about this conversation many times over the past days, and she'd decided to be strong for Moctu's sake. "Moctu...I have some bad...I have some ba—" her voice cracked and her face contorted. A tear ran down her cheek which she quickly wiped away. "He's..." Nuri fought the feeling, but it was like the ice on the rivers breaking with the spring thaw. A low moan escaped her, followed by a quiet sob which she tried to stifle. She shook her head, trying to stay strong, but then she broke down and wept uncontrollably, her face buried in the furs at Moctu's chest.

Chapter Seven

Elka could only vaguely remember it. The warrior had blackened eye sockets that made his face look like a corpse or something from the Spirit world. He'd thrown his spear which she dodged, but it sliced her shoulder. Then he'd come at her with the cudgel, and she'd turned and crashed her way through the brush and around boulders. Now she was running, running in panic, fear, and desolation.

I've failed, she thought. Lost everything. Oh, Spirits! Misha—oh, my precious, precious girl. And I left everything—all my food, my fire-starting supplies, the sleeping fur—everything. She tightened her grip on her atlatl and the one shaft. Well, not quite everything.

Farther on, her thoughts returned to Misha, and she was overwhelmed with gloom and remorse. She slowed to a stop on a lightly forested hill with large, dark boulders. As her heartbeat slowed, Elka stared vacantly at a line of ants carrying bits of leaves by the side of the trail. She scowled and shook her head, then rubbed her forehead. It was me…I did it. I sent Misha to her death…and then I ran. Like a scared rabbit. I ran from that horrible…

The pain from her shoulder brought her out of her despair.

Whew—don't feel good. My shoulder aches. Elka touched the area of the wound and rubbed it gently, checking the amount of blood. Don't feel good…weak…haven't lost that much blood. Was it poisoned? Need to wash it.

Before going on, she listened intently and watched her back trail for a long moment. She saw nothing, and all she could hear was the rustling of leaves. He's not following anymore, or at least he's not anywhere close.

The wind began to blow harder, and the upper branches of the nearby trees began to bend and sway. Elka looked to the sky and watched the gray, cumulus clouds rolling in from the northwest. The breeze cooled the sweat on her neck, and she shivered. Going to rain, she thought. No furs or gear, better find some kind of shelter. Going to be a cold night. But first, gotta wash this shoulder—killing me.

At a fast-running stream, Elka stripped off her tunic and looked at the wound. Most of the bleeding had stopped. Rivulets of dried blood ran to the bend in her arm and had caked on her skin and clothing from her shoulder blade all the way down her side. After pouring handfuls of cold water over the wound and washing off the dried blood, the cut reopened, and bright red blood flowed down her bronze-colored arm. The area around the wound was red and hot, and Elka knew that the spear had been either poisoned or contaminated with feces—or both. Lehoy had told the Nereans that, in preparation for war or hunts, Lion warriors dipped their spears and shafts in a poisonous concoction made from yew seeds and leaves. Nereans had adopted the practice when expecting battle with Lion warriors. Simple grinding of the plant material, mixed with spit was all that was needed. The poison was strongest now, in the cold season.

Elka's head swam, and she sat and covered her face with her hands, groaning softly. She couldn't tell if her weakness was from the poison or the desolation that she felt. The cold water that she'd rinsed her wound with had mostly dried, and she shook with chills as she reclothed herself. Elka gasped as her back spasmed, and she knelt and retched violently.

Oh, Spirits, she thought as another wave of nausea swept her. Strands of saliva hung from her mouth, and she panted, knees and hands on the ground, as she waited for the next wave to come.

Her thoughts were erratic and fitful. "So cold," she murmured, "…need to cover myself…cover with leaves…freeze to death." Another surge of nausea hit, and she heaved, but nothing came out. "Oh, Spirits! Grief…I can't…"

She shook her head, spitting off the strands of vomit and saliva. She breathed heavily, and through her half-closed eyes, the ground looked distorted and hazy.

Not going to make...she began the thought but shook it off angrily.

"So cold...just...need to lie down," Elka murmured again. She felt the cold ground against her face. "Just for a little..."

Chapter Eight

"HELLO, INDAR," MOCTU SAID to the tall boy with deep-set, dark eyes, who had just put his head beyond the furs at the hearth entrance. "How is your mother, Avi?"

"Mother is fine, and the baby's coming soon. Thank you, sir. I have important news—exciting news."

Moctu was seated, leaning back against sleeping furs he had propped against the rock wall near the fire. He winced at the pain in his ribs as he readjusted his position to face the young man. He forced a smile, even though the boy looked exactly like Jondu, his father, who had been, more often than not, Moctu's adversary. But Jondu had been killed by the Shiv long ago, and Indar was a hard worker, so Moctu tried not to let it affect his opinion of the boy.

"Exciting news?"

"Yes, our scouts have seen and met with a large group of boat people—people in boats—many boats. They're three days west of here," he gushed. "They don't speak anything like our language, but we've been able to use signing. They came from far away across big seas, and they've been here a while fixing two of their boats. They want to trade for meat, nuts and other food. And for tar and pitch."

Moctu was intrigued. ""What kind of boats? How many people? What do they have to trade?"

"They showed us one of the boats. It's got a wooden…um, skeleton—a wooden frame—with hides stretched over it."

"How many people?"

"Seven hands, maybe a few more." Indar tried to sound matter of fact, but his excitement was evident, and it was contagious.

"This is very interesting…and exciting," Moctu allowed. He thought for a long moment, weighing his desperate need to search for Elka and the attackers, compared with a much easier trip to the western coast. But he had to know if these newcomers were a threat, and if not, what they had to trade. "I need to make a trip to see these people and their boats," he said finally.

"That might not be wise in your um…" Indar cleared his throat, then said, "your condition." He went on quickly. "And it may not be necessary—they're sending several scouts and traders this way."

"Any word from our hunters?" Moctu asked. He worried that anyone seeing how thinly guarded Etseh was might be tempted to attack. And if the hunt had failed again, his people had little of value to trade.

"No word—nothing."

"We can't spare anyone to go search for them," Moctu muttered, mostly to himself. "Indar, tell the northern lookout to set two more signal fires. Maybe our hunters will see the signals this time. We need them back here."

Chapter Nine

Elka knew she was dying—freezing to death—but she had no energy. She was wet from the rain that had awakened her, and she was cold—so cold. Her thoughts were fleeting, fragmented and confused.

Need to get up…find a warmer… Bury myself in leaves. Yes. That's what Moctu… Just need to rest a little more.

She dreamt, and even in her dreams she knew the end was near. It was a cold, snowy winter, and the wind was blowing so hard that the snow kept hitting her face. She was chasing a buck that she'd wounded, but her body shivered uncontrollably, and her back continually spasmed and seized. She was exhausted, and her hands were blocks of ice that she couldn't move or feel.

So cold. Must get warm. Have to… If I can just bring down that buck, I can cut him open and stick my hands inside. Just need to get them warm.

But the buck stayed ahead, always just ahead, a little out of reach.

Must rest. Have to stop. Just lie here a moment. What's this? Getting warmer…better now. But… What did Nindai tell us about… Before people freeze, they get…warm feelings of peace. Is that…?

Chapter Ten

"PALO…AND KOBY…THEY'RE BACK," INDAR yelled from his lookout. "And they've got a loaded travois!"

People rushed to meet the returning hunters—both to see what was on the travois and to tell them of the deadly assault. The triumphant smiles that Palo and Koby sported changed abruptly as they learned of the murderous raid and the deaths of Meko and Ender, as well as the wounding of Moctu. Koby covered his eyes with his hand while walking a few paces away, trying to control his emotions.

Palo hugged his mate, Siduri, who carried their second boy at her hip. "I'm so glad you and Jaxon are okay. What about Luca? What about Nuri and Elka?"

"Luca's fine," she said of their oldest son. "And so are Nuri and Elka. Well, about Elka, she went after the killers, so we're hoping—"

"She went after the killers?" Palo asked incredulously, and Koby drew nearer to better hear the conversation. "Who went with her?"

Siduri just shook her head. "Misha—that's all…she had her wolf."

"One girl against two warriors? Well, I'm going then," Palo said firmly. "Even Elka is no match for two Lion warriors."

"I'm going too," said Koby. "Let's get this meat in, and we'll be off." To the crowd he said, "This is two of them—two caribou. We killed four, as well as a boar. So, two more travois are coming."

A muted cheer went up, and Siduri said, "Moctu will be pleased—finally, a good hunt. We have traders coming from the west—Sea People in boats—a large group."

"Sea People?" Palo asked. "How big of a large group?"

"About six or seven hands, almost all men, and at least two traders are coming here in a day or two—with lots of goods and things we haven't seen, so your good hunting—all this meat—will be valuable. You should stay for that long—won't you? I want you to stay for a while," Siduri said, hugging him suggestively.

Palo wanted to stay with Siduri and see these Sea People and trade, but his lower jaw was set.

"I have to leave tomorrow. Moctu's wounded and can't help his own daughter. Elka shouldn't be out there by herself."

Chapter Eleven

Elka's face twitched as Nuri rubbed a warm, wet cloth upward across her cheek. Elka grimaced and moved her head to the side trying to avoid the rasping cloth, and her eyes fluttered open. They widened, and her mouth gaped open as she realized that it wasn't Nuri after all. It was Misha, and she was curled up next to her, keeping Elka warm.

Is this a dream? Elka thought as her heart soared. Misha adjusted her position, and intensified her licking, encouraged that Elka was awake and stirring.

"Oh, Misha," Elka murmured as a tear rolled down her cheek. A warmth spread through her chest, and she whispered, "Misha, oh, my sweet..." Her elation waned as she saw the wolf's crushed left eye, still with fresh, bloody drainage alongside black, caked residue that matted her fur. Elka attempted to lift her arm to stroke the wolf and check her wounds, but her arm barely moved.

"So weak!" she wheezed, but the alarm at her own weakness was swamped by her relief at seeing Misha alive. "Thank you Spirits," she whispered, and let her body sink back toward the warmth of the wolf.

She kept me alive, Elka realized. I would have frozen. Spirits, thank you, thank you, *thank* you. With difficulty, Elka managed to raise her left arm so she could stroke Misha's fur. But her strength gave out, and her hand flopped down, hitting the wolf's back and left side, and Misha yelped in pain.

"Oh, no—I'm sorry Misha—I'm sorry." Her hand slipped down, stroking the wolf's right side, and Misha calmed, and resumed her licking.

"Wow, aren't we a pair? Two badly wounded girls. But we're still here, and we're together." Elka couldn't restrain a thin smile, in spite of their difficulties. "What are we going to do, girl? At least my head's cleared some, even if my body's still got problems."

It was a cool, dewy morning, and the fresh breeze, and the birdsong and other noises of nature began to have the same effect it always did. Elka felt the stream of life running through her veins, the same stream that ran through Misha, the birds, and even the numberless blades of grass at her side.

"We're going to make it, Misha. You're down to one eye, and I can barely move, but we're together, and we're going to make it."

The radiating ache in her right shoulder and her thirst and hunger tempered the burst of optimism. With her left hand, Elka touched the wound and was rewarded with a surge of intense pain that took her breath away.

Poison, she thought, and infected. Need to wash it again. She closed her eyes and exhaled at the thought of the pain that would bring. Elka attempted to get to her feet, but could only manage to reach a sitting position. She rolled her eyes and muttered, "Going to have to…" She looked down at the ground and shook her head. "Crap, going to have to crawl to the creek."

Elka was breathing heavily as she reached the creek. She drank from the fast-flowing, cold, clear water and murmured, "That's a little better."

Misha followed her, limping noticeably, then began lapping water at a small bend in the stream. Elka struggled out of her tunic and began pouring handfuls of water over the wound which was red, swollen and pouting. Each handful brought a new wave of pain, and Elka's head swam.

She lay back on the cold ground gathering her strength for a few more handfuls, and Misha came to her side.

Washed the poison away, she thought. But it's infected, and it's spreading. What would our healer, Nindai do? What potion…what poultice would he put on it?

Elka closed her eyes, thinking, and she took a deep breath readying herself for another few handfuls of the clean water. She was jolted by an explosion of pain as Misha began to lick the wound.

Elka swung away from the wolf crying, "Ai-ow! No Misha, No." She watched Misha back away, her ears flattened, as Elka steadied her breathing and regained her composure.

"That's all right, Misha. You were trying to help," Elka said reassuringly.

After a moment of reflection, she thought, hmm—wolves heal themselves by licking their wounds, I wonder…

Elka touched her wolf tooth amulet, nodded to herself and said, "Come here, Misha, come on." She motioned to the wolf and offered up her shoulder to the beast, and Misha obliged. Each lick was a wasp sting of pain, and Elka clenched her teeth on the leather of her tunic to withstand it. The licking went on and on, and the individual stabs of pain merged until there was a continuous mountain of it. But Elka began to feel a numbness replace some of the pain, and she hoped that meant the infection might be lessening. Misha was still licking as Elka blacked out.

Chapter Twelve

"Thanks for going, Palo—I wish I could," Moctu said. "I'm really worried about her."

"You can still barely walk, and your arm…I'm just glad you survived the attack. We'll get those scumbags who did this." Palo's scowl was replaced by a tight smile. "If Elka hasn't already."

Moctu smiled at the comment, but his brow wrinkled. "Just bring her back safe. And thanks for taking Koby. He wants to help, and it'll be good experience for him. And it's best for there to be two of you out there."

"He's a good boy…young man. He'll be an asset." Palo patted Moctu's good shoulder. "He takes after his father."

"Thanks Palo."

There was a commotion outside the hearth entrance, and they heard Indar shout, "They're here—the Sea People—the traders are here." Indar stuck his face past the hide that covered the entrance. "Two of them," he gushed. "With big packs and a travois."

Intrigued, both Moctu and Palo ventured out toward where a crowd had gathered around the newcomers.

"Moctu—he wants you. He's asking to speak with our leader," Alta called to her son. "I can't understand anything else he's saying," she said, gesturing at the broad-shouldered, shorter one of the two strangers.

Moctu slowed his gait to be able to walk with less of a limp. Both of the two traders were stocky with muscular chests and arms. Both had full beards and dark, weather-beaten skin. The shorter man that Alta was pointing at had a well-made, cream-colored tunic handsomely decorated with small, dark shells and ivory. The gray in his beard and hair suggested he was considerably older than the taller man. The men smiled at him, and Moctu saw the taller man was missing an upper front tooth.

If I could sign with the Pale Ones enough to understand them and learn their language, I ought to be able to deal with these guys, Moctu thought as he neared them. He smiled and put his fist to his heart and bowed his head. Both traders followed suit, and Moctu pointed to his hearth and motioned for them to follow him.

Addressing the crowd, Moctu said, "No one is to touch their packs or travois. Parents, please see that your children understand that. We're building trust here, so let's be respectful."

He caught Nindai's eye and said, "Nindai, you, Palo and Nuri come with us. Indar, you watch over their packs and travois and make sure nothing is disturbed." He smiled again at the traders and nodded his head toward his hearth and turned and began walking with as much grace as he could muster.

Nindai and Nuri are good choices, Moctu thought, satisfied. Both of them learned to sign with the Pale Ones, and that'll help. And Palo, well, he's my best friend.

Within the hearth, Moctu offered the most comfortable seating to the newcomers, then busied himself with restoking the fire. Soon, although cramped, the hearth was alight, warm and pleasant.

Moctu turned to the traders, and touching a finger to his chest, said, "Moctu." He then pointed to them questioningly.

The older man smiled, put his hand to his chest and said, "Arno-pey-latzu-tok."

Moctu's head involuntarily jolted back slightly, and he said uncertainly, "Arno-pey…?"

The man chuckled and said, "Arno," while nodding his head, indicating the shorter version would be fine.

The other man said, "Zeke-la-sek-nosh-tok," slurring some of the sounds through his missing tooth. He smiled at Moctu's confusion, and followed up with, "Zeke."

Zeke pointed at Arno and held up one finger, indicating Arno was the leader of the Sea People. He then touched his own chest, and raising his chin higher, he held up two fingers indicating he was next in command.

The others in the hearth introduced themselves, and Moctu frowned as he noted that both of the newcomers were transfixed by Nuri, paying her more attention than seemed proper.

Palo stood, and smiling, touched his chest and nodded to Arno and Zeke in turn. He pointed outside and said, "I have to go. I have an important effort to um…" He shot a look at Moctu who nodded in understanding. "To look into."

Over the course of the afternoon, Moctu learned that the boat people had come from the southeast along the coast after crossing big waters even farther to the southeast. Moctu was excited to hear that during their travels along the coast, they'd encountered some of his kinfolk—led by Ordu's son, Marco—far to the south. It sounded like that group was doing well.

The Sea People had been on the move for almost four moons, and two of their more than three hands of boats had ground their bottoms on rocks and needed repair. These boats were their lives—more important than anything else. The traders had seal skins, but they needed birch bark tar, pine resin and/or bitumen, all of which Nereans could provide.

Moctu smiled, thinking of the bitumen seep that was two days to the south. Nodding and spreading his arms far apart, he said, "Bitumen—we can provide a lot of that," he said, "and skins full of pine resin. He pointed at some samples of birch bark the traders had brought, and he held his thumb and forefinger slightly apart. "We don't have as much birch bark, but we can get you some of that too," he said.

"And meat," Arno added patting his stomach. With a grimace he indicated they were tired of fish and seal meat.

"And women," Zeke said finally, casting a glance at Nuri.

"Hmm," Moctu said with a frown, drawing back a bit. After a moment, he smiled, put his hands to the top of his head, fingers splayed out like antlers, and said, "We have some fresh caribou meat you'll like. But we have no women to trade."

He stood and pulled the hearth entrance hides aside. A cool breeze blew in, freshening the stale air within the enclosure. The skies were a purplish blue with no clouds, and although not yet dark, Moctu could see the first stars twinkling above the hills to the west.

Turning to Arno and Zeke, he rubbed his stomach and said, "I hope you'll eat with us now, and you're welcome to stay with us tonight."

Chapter Thirteen

The sun had just set, and the western sky still had a pink glow. A half-moon was already visible through a break in the eastern hills. The early stars had arrived to watch the daylight die. Elka was ravenous, and she could tell that Misha was hungry too.

"My head's clearing, but I have to have something to eat," she muttered. Elka tried to stand, but she was still too weak. She rubbed her aching forehead and immediately scowled as her hand brushed over her thick eyebrow ridges.

Spirits, I hate them. All this misery—all this horrible misery because I'm different. My eyebrows, my nose, my shoulders…I hate them. They're all too big.

Elka blew out a breath as her head began to ache even more. A growing despondency was looming.

The sadness was coming again.

All this bloodshed, including Ender dying and maybe father and Meko too—is it my fault? Am I really the cause of the horrible winter cough that swept the tribe? And the poor hunting?

She shook her head fiercely to resist the sadness. The quick movement hurt, but the pain seemed to help drive the sadness into the background.

Countless times, Elka had sensed the scorn and resentment in some Nereans, or she had experienced it directly. She'd overheard the side comments and too often seen people stop talking as she passed by. The worries about the curse she carried and her being different were almost certainly the reasons she was still unmated even though her menses had come the previous winter.

And then there was the time more than a moon ago when she awakened in the darkness of the hearth to the sound of Moctu and

Nuri arguing in hushed tones. What he had whispered to Nuri had doused the spark inside her.

"I asked, and he wouldn't have her!" Moctu hissed, then murmured in lower tones that Elka could no longer make out.

Elka pushed hard on her temples which helped slow the throbbing, and she powered on, wanting to finish her thoughts.

I think I'm attractive—all my relatives say so, but of course they'd say that. But even others, like Palo and Nindai, they say it, too, so maybe it's true. Palo really likes my blue eyes—talks about them every time he sees me.

And I'm a hard worker and a good hunter, and I'm healthy and I've got all my teeth. So why...why won't anybody, any man...?

She heaved a sigh trying to rid herself of that relentless question. Her thoughts were tangled, and her frown deepened.

It's not that I want to be with a man that much. It's just...just that I want someone, some man to want me. Is that weird...is that wrong? And I'm tired of being the outsider—shunned by my own people.

If Moctu wasn't my father, I think I'd have been driven out of the tribe by now. But it's not me! I'm not causing it—the sicknesses, the bad hunts. Even when I spend moons and moons away, staying with Grandmother Rah and the Krog, the Nerean hunts don't go any better. Oh Spirits, I'm just so tired. Maybe I should just go to Devu and get it over with.

As Elka scowled, rejecting that thought, Misha nuzzled her hand hoping for food or attention.

"Time to focus," Elka scolded herself.

She softly patted Misha's uninjured side.

"I know girl, I'm hungry too." After a crawl down to the creek, both Elka and the wolf had a long drink. A much harder crawl uphill brought her to where she'd seen some clover, and she ate a handful of it. It was very unsatisfying, but it was better than nothing.

Elka remembered something from Mother Nuri's account of her time alone following her escape from the Shiv. With difficulty, Elka crawled uphill to where a midsized white pine stood. Her head ached, and she was breathing heavily by the time she got there. With her knife, she managed to strip off a hand's breadth of the outer bark, exposing the white, inner bark. She carved off a hunk of the material and began chewing the tasteless, fibrous matter. It wasn't much better than the clover, but Nuri believed the bark had kept her alive during her long-ago hardship running from the Shiv cannibals.

Elka noticed that Misha was nowhere around, and she began shivering as the frigid air settled on her like a thin sheet of ice. The growing darkness matched her mood.

"Misha?" she called loudly. Almost immediately, a far-away howl of a wolf was answered by another, and she thought that the second one might have been Misha. Elka touched her amulet and bit her lip.

"Oh, this isn't good. This is all bad. So many bad…" She knew that a pack of wolves would make quick work of her. They might even kill and eat Misha, if they found her. Many times, around the campfire Elka had heard the story of a wolf pack attack that ended with the wolves savaging and eating a wounded member of their own pack.

Alternatively, if the pack accepted her, Misha might choose this time to go be with her kind. And if Misha didn't come back tonight, Elka knew she'd likely freeze. Then there was the infection that was racking her body and might kill her. If she didn't get better soon, she'd starve. And lurking out there somewhere was a warrior who probably still hunted her.

Especially during her sadness times, Elka often considered how her end might come. Now, all her deepest fears were crawling just outside the bounds of her vision, brushing against her side, blowing past her ear, whispering the countless ways it could happen.

"Oh, Spirits," Elka prayed as she held her wolf-tooth amulet and tried not to shake so violently. "Please bring Misha back to me, and let us make it safely through the night."

She stared up at the dusky sky that so often brought her peace when she was overwhelmed. She could lose herself—and her troubles—gazing at the countless stars, some in their milky pathway, some marking the well-known outlines of the Warrior, the Bull or the Seven Sisters. Somehow, they gave her a sense of unity, of purpose, and confidence that there really were spirits, and they knew what they were doing.

But this evening, her worries crushed those feelings, routing them completely. Despair swept her, and Elka realized she was being stalked not only by a murdering warrior and probably a wolf pack, but by her sadness as well.

Spirits, please help me," she prayed again.

"What would Mother Nuri do?" Elka asked herself, squeezing her forehead. "Think. Got to focus. Might be alone tonight, and it's going to be another cold one. And those wolves..."

She remembered that closer to the stream there were a group of boulders that formed a three-sided enclosure where leaves had gathered.

"I can cover myself with the leaves, and the boulders will keep the wolves from coming in more than one direction."

Before moving downhill, Elka cut a straight sapling and stripped off the smaller side branches. After carving a sharp point, she scowled at her crude spear. Won't keep a pack of wolves away for long, she thought. She looked up and was lost in contemplation for a moment, the scowl leaving her face.

Well, I'll get to meet my birth mother maybe. Moctu's told me so many stories about Effie, it seems like I already know her. Whether I freeze tonight or the wolves get me, at least I'll get to see Effie.

Elka shrugged. "That's at least something," she muttered.

The crawl downhill was easy, and Elka soon found herself wedged in the small rock enclosure with some dry leaves below her and piled around and over her legs. But the rocks she leaned against were cold, and even though Elka had removed all the wet leaves before spreading the dry ones, the ground was still moist and chilly.

"Cold as ever," she muttered as her teeth chattered. "Misha, where are you? So cold...Spirits—so hungry...so tired."

She awakened to the noise of a large animal moving through the brush near her. Her heart raced as she shrank as far back against the rocks as possible. Two wolves howled—much closer now, and her pulse quickened even more. Elka clenched tightly to her crude, sharpened stick, and in the velvety darkness, vaguely lit by the countless stars and half-moon above, she could see its yellowish white, pointed end bobbing erratically.

Cave lion? Wolf? Saber-tooth? Elka worried that her heart and pulse were beating so hard that the noise would give her away.

Whatever it is out there, it's big, and it's coming closer.

There was a snort at the opening to her little rock enclosure, and Elka's heart soared with relief. I recognize that snort—that's Misha!

Elka let out a small whoop of joy and said, "Misha, here—oh my sweet girl Misha. Here I am."

Misha entered the small enclosure and threshed about, squirming with uncontained happiness.

"Oh, Misha, Misha, I'm so glad to see you," Elka said while stroking the uninjured side of the wolf. After several moments of happy reunion, Misha drew back, her head lowered, and she began choking distressingly.

"Oh, Spirits," Elka said, "What's…" She got no farther before Misha coughed up a large, slime-covered mouse that fell at leaves covering Elka's lap. In the dim light, Elka saw the wolf back up a pace, then put her head lower, obviously proud of her offering.

"For me?" Elka was at once disgusted and honored. Misha took a half step forward, then backed up again. "Good girl, Misha. Thank you so much," Elka said, petting the wolf. She picked up the dead mouse and was immediately repulsed. It was wet, warm and slippery with slime, and it had an acrid scent of digestive juices.

Still, it was food, rich food, and Elka was ravenous. Leaving the enclosure she crawled, mouse in hand, to the creek and washed the goo and smell off its fur. With her knife, Elka gutted the animal and stripped off most of its fur. She rinsed the mouse again and gingerly took a bite of the meat near its shoulder.

Oh, so good.

Dawn was nearing by the time she made it back to her enclosure. With Misha huddled next to her, Elka felt a warmth and renewed sense of hope. "Wow, I just ate a raw mouse and sucked the brains out of its split skull," she murmured in disbelief. "One of my favorite meals ever."

That's when they heard a nearby wolf howl, and it was answered by one even closer.

Chapter Fourteen

"MMM-MMM," ARNO EXTENDED HIS hands and spread his fingers smiling broadly. "Ver good stew caribou."

Between signing and a half-day of conversing, all were beginning to better understand one another.

"Yes, fresh caribou is hard to beat," Moctu replied nodding. "We're glad you liked it. Nuri is an amazing cook and good at many other things as well. He winked at her and she smiled back at him. "Tomorrow will be a big day. I'm very interested in this bow you say can throw a shaft far and fast."

Zeke cleared his throat and mostly by signing said, "I'm the best bowman in our group. I will show."

"If it's as you say, it'll be valuable to us. Good for hunting and um…we have some difficulty with neighbors," Moctu said, unconsciously rubbing his wounded left arm. "But now is time for rest. You're welcome to stay in our hearth over there," he said pointing. "By the fire."

Still seated, Zeke slipped closer to Moctu and whispered, "Is good I use your woman tonight?"

Moctu drew back from him frowning.

Zeke whispered more urgently, "Our group—mostly men. Been long time. Too long. I give bow—no trade. Teach good. Kill many caribou…or enemies."

Still frowning, Moctu cast a glance at Nuri, who peered back at him, her head cocked inquisitively.

"Um, that won't…" Moctu stammered, still taken aback by the request. "No, that's…not our custom." He shook his head. "No," he said finally. Moctu drew himself up a little higher, and his voice rose. "Nuri speaks for herself, but I'll tell you, the last man to take her besides me ended up with his throat cut." Moctu knew

Zeke wouldn't understand all his words, but he underscored his last few words with a finger moving fast along his throat.

Zeke recoiled at the gesture, his eyes narrowing. He opened his mouth to respond, but Arno put his hand on the younger man's shoulder and spoke a torrent of words, none of which Moctu understood. The words at first angered Zeke, but then he calmed, and Moctu smiled at him, attempting to further defuse the situation.

There was no more talking, and all retired to their sleeping furs. Nuri nestled down next to Moctu and whispered, "What was that about? What did Zeke want?"

"You," Moctu replied. "He asked to 'use' you tonight."

Nuri's lip curled, and she twisted toward where Zeke lay, but Moctu held her in place.

"Easy, easy. He knows the answer's no, and the matter's settled. I told him you'd killed the last man that…tried that."

Somewhat mollified, Nuri said, "Thank you," and she snuggled closer to Moctu's unhurt side.

"That bow…the thing that throws a dart faster than an atlatl—that sounds important. Are we going to trade for it?"

"I agree," Moctu whispered, "that's the most important thing I heard today. I hope what they're saying is true."

Nuri murmured her agreement and then pressed herself into him seductively. "Want to let Zeke hear what he's missing?"

"Hmm, that sounds a little cruel," Moctu whispered, and Nuri stifled a giggle. He pulled Nuri closer still. "So, yes."

Chapter Fifteen

MISHA LET OUT A low growl and faced the opening to the rock enclosure. Her teeth were bared, and her body was crouched, braced for action. The wolf pack was closing in.

Elka, too, readied her makeshift spear and drew herself as tightly as possible into the shallow hollow between the two large boulders at the back.

This'll be over quickly, Elka thought, shaking her head. No real weapons, and both of us sick or wounded. No way we can hold off a wolf pack.

Her pulse quickened as she saw a dark figure rush by the enclosure opening, and Misha's growl deepened into a threatening snarl. The fur on her back rose, and her tail lowered.

Their pack's scouts are checking us out, she thought. She saw the flash of Misha's fully bared teeth in the predawn murkiness, and her snarl reverberated against the rocks of the tiny enclosure.

"Never seen her like this," Elka murmured, blowing out a breath. "Even wounded and one-eyed, I wouldn't want to be the wolf to face her."

A huge, dark wolf darted into the enclosure and Misha savagely chased it out. But in short order, more of them made it in—at least two. In the dim light, Elka could only see turbulent, swift movements of dark-furred beasts. Lots of growling and snapping gradually forced Misha backward, her tail almost touching Elka. A massive wolf lunged at Misha, and the two tumbled on the ground, rolling, snarling and biting. The savage growling swelled into an unceasing but shifting roar, with occasional yelps. Elka bit her lip and stared helplessly, waving her pathetic stick in front of her.

Elka heard a yelp from the bottom of the melee, and she was sure it was Misha's. She tried to get to her feet to join the fight and stab one of the wolves on top, but she was too weak.

"Going to be over soon," Elka murmured resignedly. One of the topmost wolves involved in the fight backed up close enough to her position for her to stab at it. She connected, and the wolf yelped, but the stab had no force behind it, and now the wolf turned on her. Elka could see the hideous canine fangs coming at her, and she stabbed again, but the wolf easily grabbed the crude spear in its teeth and ripped it from her grip.

Elka withdrew as far into the niche between the rocks as she could, but she was still almost fully exposed.

"This is it," she muttered, her lips pursed.

That's when another wolf joined the fray, and the wolf coming at Elka dropped the spear and turned. Morning had melted the darkness, and Elka saw the huge, new intruder throw itself into the turbulent brawl of wolf bodies, snarling and biting. Two indistinct shapes broke away and fled leaving one on the ground. The wolf closest to Elka turned, snarling at the newcomer. But after a brief faceoff, it lowered its head, its ears drawn back, and slinked slowly and carefully past the bigger wolf.

Elka gasped as she saw the white tuft of hair on the forehead of newly arrived wolf. "Loebo," she cried. Loebo studied her for a moment, but came no closer, instead turning to attend to his sister, Misha. The she-wolf lay on her side, still alive, but panting heavily and bleeding on her neck, shoulder and haunch. Loebo sniffed her neck and shoulder area, then focused on her scabbed-over, missing left eye. He licked it once, then moved to her shoulder and licked it clean.

Worried about Misha, Elka crawled closer. As she neared the two wolves, Loebo turned toward her and showed his upper teeth. Although he didn't growl, the threat registered, and Elka moved no closer.

"It's me, Loebo," Elka said in a calm voice. "You remember me, don't you? You've gotten so big, Loebo. Are you the pack leader? Where's Peet? Where's your brother?" She hoped Loebo would remember her voice and be soothed by it. Loebo continued to lick his wounded sister and didn't seem to mind as Elka once again edged closer. "How's Misha?"

At the mention of his sister's name, Loebo turned to Elka and appeared to better remember her. He stared at her for a long moment, sniffed once, then went back to licking Misha's wounds. Elka moved close enough to touch Misha's leg and could see her wounds were severe—especially the haunch. She softly stroked the fur above the wound and said, "My poor, brave Misha."

In the dawn light Elka could see the pack—easily six or seven of them—milling a short stone's throw farther into the forest.

Hmm, this may not be over yet, she thought worriedly. But her worries melted into relief as Misha rolled to a position lying on her stomach, her feet under her.

Chapter Sixteen

Much of the trading had been accomplished before the sun began its decline. Although reluctant to give up any of their precious, limited stores of fresh and dried caribou meat, Moctu and the Nereans had received some superbly processed seal skins. Moreover, for the meat and for skins full of pine tar and bitumen, they'd gotten three fine bows with their accompanying arrows.

"Spirits, they do send the shaft far and fast," Moctu said. "Much better than an atlatl. These'll help with our hunting and with…well, some of our other problems."

Arno smiled broadly, proud of their technology, and he patted Moctu on the shoulder after one of his shots hit the target stump. He pointed at Moctu and said, "You fast…" He nodded then shrugged and jiggled his hands in front of him, searching for a word. "Good, good. Even with hurt side." He held up two fingers. "Two summers be ver good."

"We can make more of these," Moctu said enthusiastically, pointing at the bow. We'll do as you said. We have the yew wood, and we'll find some with no knots. And the heart wood goes to the inside, and the younger, sapwood goes to the outside."

Arno smiled again, and Moctu found that he really liked the man. As for Zeke, not so much. The young man had gone south with a group of Nereans to mine the bitumen the Sea People needed to patch their boats. Indar had gone, too, but the rest of the group were women, including Leuna. She was Seetu's mate, and he had yet to return from the hunting trip.

Along the trip down to the bitumen seep, Zeke became surly after making unsuccessful passes at all the women individually. On the second day at the seep, he cornered Leuna as she knelt, scraping up bitumen residue near a black, oily pond where putrid bubbles swelled to the surface and burst. As he grabbed her from behind and fondled her breasts and bottom, Leuna's screams brought Indar, and the two were able to fight off the larger man. The group began their return trip that afternoon, and the Nereans stayed close together with Zeke isolated, walking some distance behind.

Zeke had kept to himself since the return, although he'd been the first to demonstrate the bow, And, wow, had he been good. His first three shots had not only hit the target stump, but they'd been closely spaced near its center. Arno was an excellent shot as well.

Moctu cleared his throat. He'd saved one final piece of trade bait to the end. He wanted more bows and strings. His people had been instructed to protect and not show their fire-making technique to either of the traders. Now, he was going to find out what he could get in trade for the procedure.

"What would it be worth..." He frowned, knowing they couldn't understand those words. "We can make fire. What will you give to learn?"

Arno smiled, patted his shoulder again and nodded dismissively. "We know...we have the rocks."

Moctu's brow furrowed, and he cocked his head in confusion. "The rocks?"

"Flint and gold rock," Arno said. Then he looked quizzically at Moctu and asked, "How you make fire?"

"We use wood. Maybe we just trade to learn each other's way?"

Arno didn't understand, so Moctu said, "We show how we make fire, you show how you make fire—we trade."

Arno nodded then went to his pack and withdrew a sealskin pouch. He pointed to a place with less grass which was shielded from the breeze. After quickly building a semicircle of rocks, he took a piece of cattail from the pouch and with his thumbs, fluffed a chunk of it into a cottony mass.

Moctu nodded. He'd used the same material as tinder many times. Arno crumbled some dry grass above and around the pile, leaving some of the cattail fluff exposed. He took an unfinished flint spearhead from the pouch and a rough piece of rock with gold sparkles. Arno struck the rocks together—a glancing blow that produced a tiny shooting star of light. Moctu had seen those before while knapping spearpoints, but they weren't hot and never lasted. With just two more strikes, Arno hit a spark into the fluff, and it began to smoke. He picked up the fluff and dead grass and blew on it and showed Moctu a handful of fire. The whole process had taken less than ten heartbeats.

Moctu's jaw dropped, and he expelled a breath and began laughing.

"Spirits, all this time we've missed that," he exclaimed, shaking his head. "Those weren't tiny shooting stars. They were sparks of fire!"

He felt almost silly showing Arno his way of fire-starting by rotating a wood spindle. It took inordinately longer and left him sweating and a little out of breath. Even so, Arno was interested and moderately impressed. Both seemed surprised that there were two ways to make fire.

Arno nodded appreciatively and said, "Wood fire-making help if lose gold rock."

Still stunned and excited, Moctu smiled broadly and said, "You stay. I'll be right back." He couldn't wait to show his people the new way of making fire. It was another valuable learning. Returning quickly, he presented Arno with his best spearhead. It was beautiful—long and thin and well-crafted from colorful red and black flint.

"Gift," he said simply, and Arno beamed.

Chapter Seventeen

It was two days since the wolf pack had left, and both Elka and Misha were struggling. Elka had a fever, and she was extremely weak from lack of food. She was disoriented, her thinking once again confused. She'd stripped the nearby surrounding area of any vegetation that could provide nourishment. Misha, too, was undoubtedly ravenous, too weak to hunt and limping badly when she walked.

"Oh Misha, how are we…" Elka shook her head and clutched her stomach which lately hurt more than her wound. She mumbled, "What are we going to do?"

She dozed off and was awakened by Misha growling. But the wounded wolf didn't rise to her feet, and Elka closed her eyes and drifted off again.

"What? Aagh, no," Elka said, shaking her head and shifting her position as something cold touched her shoulder wound. Her eyes opened, and she gasped, immediately transforming from drowsy sluggishness to high anxiety. "Aagh, no…no," she squealed as she tried to squirm away from the tall, dark man who knelt as he attended her wound. But Elka didn't get far, and even that brief exertion left her drained and breathing heavily.

The man finished applying the poultice, then wrapped a strip of hide around her arm to hold it in place. Satisfied, he turned his attention to her face and eyes. His gaze lingered on her blue eyes, and he cocked his head slightly. But then he put his hands to his chest, grimaced and stuck his tongue a little way out. His brow wrinkled and he pointed at her and said, "Sick, ver sick." He got up, and as he moved away, Elka was surprised to see he had a fire going and a skin dangling from a horizontal stick above it. He was warming something.

Even in her foggy state, she wondered how he had gotten past Misha, who typically attacked, or at least threatened strangers. She twisted her head to see Misha intently chewing a meaty bone.

She was so hungry, she'd treat anyone with food as a friend, Elka mused, and she allowed herself a thin smile. Looks like a caribou bone, and that's Misha's favorite. Elka attempted to swallow, but her mouth was so dry that she couldn't finish it. She pursed her dry lips. I'm that hungry too—and thirsty. Elka studied the man, who had his back to her. Tall, she thought. Black hair pulled back and tied with a leather strip.

He had broad shoulders, narrow hips, and he was wearing well-tanned and well-sewn hides that were silvery-white with dark spots. They didn't look like caribou or any other hide that she'd seen.

He doesn't seem to want to hurt me or Misha, Elka thought, and a growing sense of relief spread through her. She felt herself relaxing and edging toward sleep again, but she fought it off. The man turned to her with a wooden bowl in his hand. He smiled that she was awake, and he knelt by her side. As his knee touched the ground, Misha released her bone, and gave a deep, menacing snarl. Elka saw the man freeze, his eyes wide with alarm. After a long moment where no one moved, Misha went back to gnawing her bone, and the man let out a breath.

He gathered himself, and lifting Elka's head, he brought the bowl to her lips and nodded his head for her to drink. Elka smelled the contents, then took a small sip. It was hot and wet, with tiny shreds of meat in it, and it was delicious. She finished the cup, and he left and returned with another helping. After Elka finished that one, he lay her head back, pointed at her, then closed his eyes and tilted his head bringing his free hand to his ear, simulating sleep. It took no more persuading than that, and she was quickly asleep.

It was a sunny morning when Elka awoke, and she found that Misha had moved to her side. At least she's moving around a little more, Elka thought, propping herself up on one elbow.

"Hmm, I am too," she murmured to herself, smiling. The sun's rays were streaming through the nearby trees, dappling the ground below. Birds were singing, and Elka felt an unrestrained buoyancy and happiness. "My stomach doesn't hurt any more, and my shoulder feels better," she marveled. "Misha seems happy, and I'm…I'm thinking clearer."

Elka saw the man, seated on a boulder, finishing a cup of his wonderful soup. He noticed her watching him, and he smiled at her pointing that she was up on her elbow. He pointed to the cup questioningly, and Elka nodded her head.

As he approached Elka, Misha raised her head and followed his movements warily, but she didn't growl. Elka pointed at him and asked, "Who are you?" She studied him as he knelt by her. His black beard was very short, indicating that he'd recently shaved. His features were rugged but pleasant, and those shoulders—they were broad, almost like a man of the Krog.

He offered her the cup of soup, and with his free hand he touched his chest and said, "Pele-shok-nu-tok." He grinned broadly at her look of consternation, then said, "Pele." Pointing questioningly at her, he cocked his head.

"Elka," she said smiling and tapping her chest with a finger. "Thank you, Pele," she said dipping her head and briefly closing her eyes. "Thank you for your kindness."

He smiled in understanding, and tipped his head. On Elka's other side Misha changed positions to face him while still remaining close to Elka. Pele pointed to the wolf and spread his hands quizzically. "How…um…how you…?"

Elka stifled a chuckle and stroked the wolf's fur. "Misha," she said. "Misha is my…my sister, my friend."

"Misha," he said, looking at the wolf warily. He looked back at Elka and questioned, "You…um…you shaman?"

"No," she shook her head and continued stroking Misha's back. "I just like wolves, and they like me. Most of them anyway."

Pele frowned, rubbed his forehead and repeated back some of her words.

"We'll come back to that," Elka said smiling. She softly patted her bandaged shoulder. "What here? What's in the poultice? Seems to have worked."

"Ah," he said and got up. He searched through his large pack and held up a handful of dried plant material.

"Oh, yarrow," Elka said recognizing immediately.

Then Pele extracted a handful of what, from a distance, looked like gravel.

Elka frowned and cocked her head, but as he neared her, she saw that it was garlic—cloves of garlic. As he knelt, she pointed at his hand and said, "Garlic, we call that garlic."

"Gallic," he mimicked. Rubbing his fingers on his palm, he indicated that he'd crushed it and applied it to the wound. Then he rubbed his stomach and pretended to drink from a cup.

"It's in the poultice, and it was in the soup too? Well, it seems to have worked." She once again touched her chest and looked down. "Thank you, Pele. I think you saved my life. And Misha's too."

Pele nodded, smiling in understanding. Then he pointed to her blue eyes and let out an appreciative "Hmm."

Elka blushed a little but was pleased by his interest.

"How um…" Pele said. He paused, then started over. Pointing to his own eye then to the cloudless sky, then again to her eyes, he said, "Ver good."

Chapter Eighteen

"Wow—you're picking it up fast," Moctu said to Lehoy after he hit the stump on two successive shots. The two men had long been the best in the tribe with atlatls, and now it looked like that would also be the case with the bows.

"You're right there with me, even with all your injuries," Lehoy responded.

Moctu eyed the tall, good-looking man with the two long braids of dark hair, since it had been members of his former tribe that committed the attack. Lehoy had once been the Lion People's best hunter, but he'd long ago fallen in love with Avi, a beautiful, green-eyed Nerean woman. That love, combined with Lehoy's dislike of Devu, the shaman of the Lion People, had driven him to join the Nerean tribe. Since then, Moctu and Lehoy had fought alongside one another against the Shiv cannibals and in skirmishes with Lion warriors. Besides Palo, Lehoy was Moctu's best friend.

"Yeah, it hurts my side to pull the bow back, but it's getting better fast." Moctu clapped Lehoy on the back and smiled. "You know, it was your successful hunt that helped us get these bows. Not only are we set with food for more than a moon," Moctu continued, "all that meat was valuable in our trade talks with the Sea People."

Lehoy beamed and nodded his head. "That was a good trade—not only the bows and the seal skins, but the fire-making with rocks. Wow—the Spirits blessed us."

"Now I just need to get Elka back, safe and well. And Palo and Koby—haven't heard from them either."

"Palo's a good man and he'll take good care of Koby. I bet we'll see them all soon."

Moctu blew out a breath. "Hope so," he said biting at a rough edge on his thumb nail. "Nuri and I are going to leave soon to look for them all. I think I'm healed up enough."

Lehoy studied the fingers of his right hand. "I think I'm done with practice for today. Getting a blister."

As the two walked back toward Etseh, Lehoy noted Moctu's limp. He pursed his lips then said, "Might give it another few days before a big journey, friend."

Back at Etseh, Moctu asked Indar and Zori to come to his hearth. Zori was the son of the former leader, Ordu, who had been brutally murdered more than a decade back. A year ago, Zori was mated to Moctu's young, half-sister, Zaila, and they were happy. Both Zori and Indar were big, energetic and dependable.

"I asked you here because you're both young and strong, and I trust your judgement," Moctu started, and the boys glanced at each other and smiled with pride. "I have a special…um, mission for you—one that'll take care and stealth."

The boys' eyes widened and they leaned in to hear his words, as his voice lowered.

"I need you to follow the Sea People traders and keep watch over their camp—all without being seen. Do you think you can do that?"

Both boys nodded vigorously, their eyes aglow. "Yes," they said in unison.

"The Sea People are our friends, and we have a good relationship with them," Moctu said. "I don't want them to think we don't trust them. But there are at least six hands of them, mostly men, and they have weapons better than ours. We want to trust them, but we need to make sure that their…their intentions stay friendly and peaceful. If you see any…" Moctu's hands were open in front of him as he searched for the right words. "Anything disturbing or…just anything that worries you, come tell me right away. And keep in mind, if they see you, they're not going to know that you're friendly, and they…" He raised his eyebrows in warning. "…they might shoot at you. And they're both really good shots."

Chapter Nineteen

Elka could stand and walk, and she was rapidly recovering her strength. Misha, too, was up and about more, and her limp was almost gone.

It's amazing what food and rest—and a little garlic—will do, Elka mused as she sat on a boulder near the fire. Once again, her gaze settled on the tall, powerful man who had saved her life. While he intently organized his pack, she studied his face. It was weatherbeaten, but handsome, and his smile was captivating. And Pele smiled at her frequently.

Conversations had become easier as each learned more of the other's language. She'd learned all about his people that came by the sea in boats. He'd had a mate, Seega, who died in childbirth two summers before. The infant, a boy, had also died. Pele's voice caught as he talked about them, and Elka wanted to hold and comfort him, but she couldn't gather the courage, and the moment passed. Pele spent much of the rest of that day by himself.

Pele was intrigued by her background, being partly a Nerean and partly from the Pale Ones—the Krog. He'd also wanted to know more about how she'd charmed a wolf into being a companion and helper. Misha remained watchful but had only growled at Pele once more when he set his huge pack down roughly.

She likes the caribou meat he's been feeding her, Elka thought. Wonder when he's going to be on his way? He's been good… It's been really nice having him around. I've talked to him more in the last few days than anyone at Etseh in the last moon. And I can tell he likes me. There aren't that many people at Etseh that…Elka shook her head and her thoughts trailed off as she stared wistfully to the southwest. Her eyes settled on a boulder near the creek while she ran her fingers through her hair.

"Oh, Spirits, it's filthy," she muttered.

Impelled to action, she had soon gathered some ashes from the edge of the fire and was testing the water in a rocky side pool of the creek. She whistled softly at the thought of immersing herself in the icy-cold water. It was late afternoon, and the day would only be getting colder.

"This is going to be a quick one," she murmured. She removed her clothes and the bandage around her shoulder and stepped into the frigid water, steeling herself for the plunge. Holding the hand with the ashes above the water, she quickly dipped the rest of her body underwater, rubbing her face, underarms and legs with her free hand. Breaking back above the water she trilled, "Brreeeee!" Using both hands she worked the ashes into her hair then rinsed it thoroughly. As she emerged from the creek she saw Pele, who had just arrived after hearing her cry. Embarrassed, Elka gathered her clothes and held them in front of her as she disappeared behind the boulder, all while Pele stared transfixed.

Pele had stoked the fire when she came back uphill, her teeth chattering violently.

He touched his chest, "Sorry, I thought trouble…your cry."

"Thanks for rebuilding the fire. I'm f-freezing," she said, holding her hands just above the flames.

"You…um…" Pele paused for a long time wondering if he should finish the sentence. "You ver pretty."

Elka blushed and looked away, embarrassed but thrilled. It was the first compliment that she'd ever gotten that didn't come from a family member or family friend.

"C-can't get warm," Elka said, putting more of herself directly over the fire.

"Here," Pele said, pulling a fur from his pack. Moving behind Elka, he wrapped it and his arms around her. The fur was soft and warm, and his embrace was comforting. Elka melted into him, her shivering replaced by a happy glow and a heat spreading through her chest.

Chapter Twenty

ALTHOUGH THE TRADERS, ARNO and Zeke, had left more than a day before, they were carrying heavy backpacks, and they alternated with one another on pulling a fully laden travois, so Indar and Zori caught up with them easily. The young men kept the traders mostly in sight and within earshot, but they were careful to remain hidden.

The traders looked tired, and they stopped in the mid-afternoon and built a fire. The boys figured at the rate they were going it was still a couple of days travel to their encampment at the sea.

"Hope you brought warm furs, because we'll have no fire tonight," Zori, the older boy, said.

Indar nodded, knowing a fire was a bad idea if they wanted to remain undetected. The boys snuck close enough to see the men, and they could hear the tone of their conversation. They didn't need to be closer, because they couldn't understand the Sea People language anyway.

"They're arguing," Zori whispered, and Indar shook his head. They watched Zeke point back the way they'd come—toward Etseh. His voice was raised and he gesticulated hotly, while Arno watched him, yawning occasionally, apparently unimpressed. That seemed to make Zeke even madder.

"Is that disturbing enough to go tell Moctu?" Indar asked in a hushed tone.

Zori grimaced and squeezed the bridge of his nose, then whispered, "No, I don't think so, do you?"

Indar shook his head no.

After eating and restoking the fire, the men bedded down. The boys ate dried caribou and moved slightly farther away, so that if either of them snored, they wouldn't alert the traders. When dawn broke, the skies were clear, but it was so cold, the boys could see their breath.

"No wonder I slept so badly," Indar complained. "Everything on me is cold. I didn't bring enough furs."

"Yeah, winter's coming early this year," Zori said. He winced while putting his hands behind his head and stretching. "Got a kink in my neck."

During the morning, the boys followed the men at a safe distance, and there were two sharp exchanges between the traders. The sun had just begun its downward decent when Zeke again pointed in the direction of Etseh, and animatedly gestured, his voice bellowing at one point. Arno sounded annoyed, and barked back at him, extending his arm, which Zeke batted away. Arno took a menacing step toward Zeke, and the men just stared at one another for several heartbeats. Finally, Zeke turned away and spit to the side. Arno cocked his head and scowled but let the insult pass.

The boys traded questioning glances, but after a pause, Zori shook his head and whispered, "We're not going back to Moctu yet."

All that day, the bright sun poured its soothing heat downward, and the shadows had just changed direction when the boys put away their furs. Now, they were thankful for the slight, cool breeze. It was the final uphill stretch before the flatlands leading to the shore about a day away. It was clear the traders were exhausted and cranky from pulling their ponderous travois and carrying their heavy packs. Their few verbal exchanges were gruff and curt.

Arno, pulling the travois, was just reaching the crest of the hill when it happened. Following a few steps behind Arno, Zeke picked up a cobble, and after furtively checking his surroundings, he moved directly behind Arno and bashed the rock against the older man's head. Arno made no sound, and went down like a skin full of moist meat. Zeke knelt and hit the fallen man again to ensure he was dead.

Wide eyed, the boys looked at one another, and Zori nodded. Pointing at Indar, he continued to nod his head and then jerked his thumb toward Etseh.

Chapter Twenty-One

"Mmm…how about you?" Pele murmured, still holding Elka, wrapped in a fur, close against his chest as they stood near the fire. "You have…" He started the question but paused, worried about the answer. "Your mate—he worries and searches for you?"

"I'm…" Elka's head sank. "I'm not mated."

"Woman pretty as you? And your eyes…ver pretty. How no mate? Was he…killed?"

"No, I've never been mated," she said simply, not wanting to explain. Pleased by his words and the sudden, brighter tone in his voice, Elka didn't want to ruin things by discussing her curse.

"Huh," he said. There was a long pause, and neither spoke.

Whew, she keeps so much inside her, Pele thought, his head tilting. *Like how she tamed a wolf, and how she and her wolf got their wounds. Her thoughts are like the moon where most of the time, much is hidden.*

Elka bit her lip and closed her eyes. *Should I just tell him? What if he…? Oh, Spirits, what do I do? It's probably better to get it over with,* she decided. Taking a deep breath, she turned to face him. His eyes were searching for hers, but she thought she saw a kindness there.

"I'm…" Elka dropped her gaze to the ground. "Well, there's a shaman, a powerful shaman, Devu, who says I'm…" She shrugged. "I'm cursed. He's never liked me—calls me Sleeps With Wolves, and he says all the problems we have are because of me. Says the Spirits hate me because I'm part Pale One. And no man has…" Elka felt her eyes water, and she shook her head fighting not to cry. "No man's wanted me." Her voice broke with the last couple of words, and she quickly wiped one eye.

Pele understood much of what she said, and he replied, "We go to shaman. I fix."

"No," Elka said a little too forcefully. "No, that won't work. Devu says I'm cursed, and he wants to kill me…to sacrifice me to get rid of the curse."

Pele had followed her words carefully, and he only half-stifled a relieved laugh. "So, this curse…is disease, or no?"

"No," Elka said, quickly wiping her other eye and looking up at him. She allowed herself a thin, hopeful smile, encouraged by Pele's lack of concern. "But the shaman blames me for our poor hunting and for the coughing that everyone got this winter."

Pele snorted and rolled his eyes. "My people have cough more bad this winter, and you not there."

Elka smiled broadly, and she couldn't resist hugging him. He hugged her back and kissed her temple, then her cheek. Her eyes widened as his lips continued toward her mouth. Her pulse quickened, and she felt a breathlessness as his soft lips gently found hers.

Elka was lost in the feeling of it, and her thoughts raced. My first kiss! This man likes me…and wants me!

Pele drew back and saw her smiling with her eyes closed, her lips slightly parted. She opened her eyes questioningly, noting an odd gleam in his eyes, and he bent to kiss her again. His tongue caressed her lips, then he kissed her with more pressure, his tongue moving in farther to find hers. She followed Pele's lead, and her tongue explored his.

Pele slowly moved to kissing her cheek then brushed his lips along her neck below the ear. Waves of chills swept through Elka, and her whole body felt light and unsteady. Pele sensed her knees buckling, and he helped her softly to the ground not far from the fire. He paused to study her face and saw her stunningly blue eyes, so recently full of tears, now sparkling as they reflected the firelight. She was so beautiful—shaking a little, but she was smiling.

He kissed her again, a deep, urgent kiss that drove his desire. "Slow down," he murmured to himself. "This is her first time."

For her part, Elka felt his manhood, hard and bulging, trying to free itself from his loin clothes, and she beamed with satisfaction that she could have that effect on a man—especially this man. She was momentarily lost in thought, trying to remember all of what Nuri told her to expect in these situations. Nuri had told her ways to fight a man off, and, on the other hand, things that men liked, and the best ways to please them.

Oh, Spirits, I want to please him. This is going to happen. I want this to happen!

Pele's kiss became more tender, then his mouth traced its way to nuzzle her ear. His soft, warm breath as he nibbled her lobe once again sent chills through her and caused her skin to tingle. His hand stroked her wet hair, then moved down along her neck and shoulder, then down farther, exploring the delicious curve of her hips. Elka reveled in his touch, thrilled by his interest in her.

Elka remembered Nuri's words: "Men, when they just want sex, take you quickly. But men, when they love you, they take you slowly, wanting to give you pleasure just as much as they want it for themselves." Elka sighed, smiling at the recognition that Pele wasn't rushing things.

Pele's warm hand slipped beneath her tunic and caressed her back, enjoying the warmth of her skin and the textures of sensuality. Moving first upward to her shoulder blade, he slowly slid his fingers down along the supple muscles of her spine. Elka's body bowed upward, and she let out a soft moan as an electric wave surged through her.

She's responding—she's enjoying this, Pele thought smiling, his confidence growing. *Her back is so strong, and her skin is so smooth!* He pressed his groin closer, wanting her to recognize the effect she was having on him. With his hand behind her, he crushed Elka to his body, and his kiss became more insistent, his tongue thrusting deeper. His hand moved to her side, then glided easily beneath her loose belt, cupping her bottom and pushing it to him.

Elka moved a leg between his, and he responded, pushing his hardness against her hip. Lightheaded with pleasure and happiness, Elka kissed him harder, softly sucking on his tongue, which he seemed to like.

Pele hand drifted upward and to her front softly gasping her breast. Her tunic was tight there, and when Elka made a mewing sound and raised both arms up, Pele took the signal and helped pull it off her.

"Oh, Spirits," he said huskily. "You beautiful." Pele pulled off his shirt, and Elka marveled at his powerful, scarred chest, which made him look even more manly. When she delicately touched a large, well-healed scar on his shoulder and looked questioningly, he explained in signs that all the scars and even the recent scratches came from killing big fish and sea beasts.

Pele hugged her to him, enjoying the heat of her breasts against his chest. He pulled away enough to gently stroke one of her full breasts, then he circled the fat nipple with a finger, finding it hard and upright with desire. Pele felt the powerful surge that ran through her body, and he stopped to look at her face, finding her eyes open. She looked serious, but happy and excited. Her face was flushed, and no longer cold, she had a strand of hair matted to her forehead. Pele was partly atop her, and he kissed Elka again, more gently than before. With her hard nipple between two fingers, he softly jiggled her breast, and her movements below him indicated she enjoyed it.

Pele's lips brushed her neck and upper chest, then he buried his face between her large breasts.

Her scent was of green meadows at sunrise and a faraway, morning campfire. He kissed her breast, and traced his tongue toward her nipple. Elka gasped as it got there, and began circling. Reflexively, Elka bowed her chest upward, and Pele took the taught nipple between his lips, sucking gently.

Spirits, my heart is beating fast, Elka thought, hugging him tighter. She was panting as Pele's mouth moved to her other nipple. As his tongue teased the second nipple, his hand moved slowly down her stomach to her loin wrap. It excited him when she sucked in her abdomen to free up space for his hand to reach beneath the wrap. His hand moved down, feeling the soft hair of her mound, then continued lower. Elka felt a yearning that she couldn't describe and unconsciously moved her knees apart. Pele and Elka let out simultaneous moans as his fingers found her wetness, and his middle one entered her.

Pele's manhood throbbed impatiently within his loin clothes, and he paused trying to regain control. Lost in her own world of wonderful, new sensations, Elka thrust her hips upward, wanting more of the luscious feelings.

"Oh, woman," Pele murmured. "I want…" He lost all focus as Elka's pelvis began to rotate. His finger found her nodule, rigid with excitement and longing.

"Pele…oh, yes…oh," Elka moaned with pleasure as his finger moved with more pressure and speed. "Oh, Spirits!"

"Aaaaaaaaaah!" Elka cried as her whole body tensed, and her hips bucked upward. "Aaaaaaaah," she cried again, grinding her pelvis against his finger, lost to the world in her personal ecstasy. Elka was breathing heavily, and her eyes were still closed as her hips sank back to the ground. Opening them, she reached a hand behind Pele's head and pulled him into a deep kiss. As he lay more atop her, she felt his hardness on her leg, and her hand moved to stroke it.

"I want you," she murmured thickly.

Pele moved backward, his mouth open, his eyes closed, relishing the feeling. Elka struggled with the tie on his wrap, and he helped her loosen it. His large member eagerly sprang from the wrap, thick and rock-hard.

Elka's eyes widened at his size, worried it might not fit, but wanting it nonetheless. The two of them fumbled her wrap from her, and Pele shifted to lie against her. He kissed her passionately, urgently, his tongue thrusting in deeply beside hers. Elka felt him shudder as she took hold of his shaft and began stroking it up and down.

"Oh, Spirits. Feel…good." He kissed her again, and was thrilled when she moved her legs apart, signaling once more that she wanted him. Positioning himself, he touched the head of his manhood to her wetness, and with his hand, he moved it in small circles, as Elka desperate for it, thrust toward him. He gasped as the head of his member entered her hot, wet, muscular opening.

Nuri told me this first time will hurt a little, but I want it so much. Spirits, do I want it, Elka thought, moving her pelvis more toward him.

Momentarily lost in his intense need, when Pele found her obstruction, he remembered this was her first time. He slowed, but Elka moved powerfully into him, and they were quickly past it. Elka made no cry and seemed to be in no pain. Freed from that concern, Pele thrust deeply into her, expelling a breath with the pure pleasure of it. Already on the edge, he held back, wanting to savor the feeling.

Pele felt Elka move more urgently, raising her pelvis to increase the penetration and impact of his thrusts. When she moaned, he knew she was once again close, and he pounded into her over and over.

"Oh, Spirits," she cried, and her whole body stiffened.

"Oh, gaaaaaah," Pele sighed, as the exquisite waves of pleasure swamped him. Already buried to the hilt, he kept pushing, unconsciously wanting it ever deeper.

For both of them, the wonderful feelings reverberated strongly, but like echoes in the canyon, they softened and gradually subsided, leaving a satisfied, affectionate closeness.

Elka lay comfortably with her head nestled between Pele's encircling arm and his chest. Their breathing had returned to normal, but Elka could hear Pele's heart was still beating faster than usual. His few languid movements told her he was in a contented drowsiness.

Elka blew out a small breath, marveling to herself. "I've never felt like that before. Never been…" She shook her head imperceptibly and smiled. "Never been this happy."

Chapter Twenty-Two

MOCTU'S MOUTH FELL OPEN, and his eyes narrowed. "Are you sure?"

"Indar nodded. "Zori and I both saw it. He sent me back here to tell you."

"Are you sure?" Moctu repeated, but waved it off and squeezed his forehead, his eyes closed tightly.

"I'm sorry Zori had to see that," Moctu said, remembering that Zori's father had been killed the same way when the boy was very young. "Sorry that either of you had to see it."

After a pause, he began to pace while staring at the ground and muttering to himself. "That probably means Zeke's in charge. Never liked that guy. Or trusted…" He shook his head not finishing the thought. "Probably blame it on us." He stopped and nodded, lost in his thoughts. His hand rubbed the short beard of his chin. "More of a threat now. Need Palo and Koby back here. Signal fires. Need to meet with the elders…call a council of warriors." Noticing Indar was still there, he said, "Thank you Indar, you and Zori did well. Please tell Nabu, Nindai and Ono that I need to see them… Now." He waved the boy off, then began pacing again.

The sun was high in the sky, but still rising, when Ono and Nindai arrived at Moctu's hearth. Nindai limped in first, using his cane to offset his lame right leg, which was about four finger-widths shorter than his left. Unable to hunt, Nindai had become an excellent healer and a powerful shaman. Ono shuffled in next, looking every bit of his age, as one of the two oldest people in the tribe. Both he and Jelli, a thin, wiry, female shaman, had lost count of their years, but each had been through about ten hands of winters. Moctu noted that Ono's short beard was completely white, and he wore a woven cap to cover an ever-growing bald spot. Recently back from a hunting trip, Nabu came in last.

"So, we have new worries," Ono said in a loud voice. "Now there's the Sea People to the west of us, and as always, there's the Lion People to the east."

Moctu nodded. "Did Indar tell you all of what he and Zori saw?"

"Yes, very troubling," Nindai said. "Especially that, as they argued, Zeke kept pointing back towards Etseh. You think he was trying to get Arno to agree to attack us?"

Moctu shrugged. "That's what it sounds like to me. And now that Zeke's in charge..." He shook his head, frowning.

"They have better weapons, or at least more of them than we do, and they're way more skilled at using the bows than we are," Ono said.

All four of the men were aware of the fact, but now they were silent, contemplating it.

Moctu broke the silence, saying, "Nuri and Avi built two signal fires yesterday, so with luck, we'll have Palo and Koby back soon. And maybe they'll bring Elka with them."

Ono cleared his throat but said nothing, and Moctu suspected he was thinking that Elka was the cause of this new misfortune.

"I think we should focus our activities on making more arrows and some bows," Moctu said. "Besides getting the rest of the caribou meat dried, making the bows and arrows—and practicing with them—those have to be our top priorities. Even if the Sea People don't come, having more practice with the bows will help us on our next hunt, and it'll help against the Lion People too. What do you think?"

Nindai had been staring at the embers of the fire, but now he looked at Moctu and nodded. Ono also nodded, murmuring, "Uh-huh, I agree."

Never one to say much, Nabu shrugged and bobbed his head in agreement.

Chapter Twenty-Three

THEY MADE LOVE TWICE more that first day, and each time it became a little less urgent and ever more tender. Two sunrises had come and gone since then, and Elka found that the smile rarely left her face.

"He's happy too," she murmured, her smile broadening as she eyed Pele collecting firewood. "He's got a…it's like a yearling antelope bouncing around. Happy and playful." Elka stifled a chuckle as Pele began humming.

Pele moved out of sight, and Elka's smile began to fade. The niggling worries began again. *Things this good can't last. We've been here for days. When is he going to have to move on? Is he going west back to his people or east for more trading? What if he wants to trade with the Lion People? Can't go there with him. And if he goes back to the Sea People—will he want to take me along? Then what happens when their journey continues—do I want to go with him? How about Misha? A wolf's not going to like life in a boat. I'm not going anywhere without Misha.*

Although they were getting better at each other's language, they hadn't addressed any of those questions. And how about her loved ones?

"I'd so miss Moctu and Mother Nuri. And, of course, Grandmothers Alta and Rah. And Palo…and Nindai and Hawk," said softly to herself. "But the others?" She frowned and shrugged.

"And it could help." She considered that, cocking one eye upward. "Maybe the fight with the Lion People would stop. Maybe everyone's hunting would improve…and the coughing…" She paused, then scowled and blew out a breath. "Don't do that again," she told herself sternly. "I'm NOT the cause of all that."

She looked up as Pele returned. He held up his unstrung bow and said, "Want hunt again?

Elka smiled and shook her head vigorously. His bow was amazing, and he said she was rapidly learning to use it. Yesterday, she had killed a duck on the far side of a large stream, and Pele had been very impressed, both with her shot and with Misha—who forded the creek and when Elka whistled, returned with the duck.

They had learned a lot from each other. Pele was fascinated watching Elka make a fire with her board and spindle, and she'd felt perhaps, somewhat prouder than she should have. All of that deflated when Pele took out his fire rocks, shredded some dried fungus, and with three hits of the rocks, he had a spark in the fungus fluff which he then blew into a flame. It was amazing. Elka couldn't wait to show Moctu and the others that, as well as the wondrous bow and what it could do.

Elka had already considered the fact that, if she'd had the bow, both of the marauders that injured her kinfolk would be dead. The second one, who charged up the hill after her, wouldn't have been able to fend off the speedier arrows, and she and Misha would never have been wounded. Although heartened by the thought, it also distressed Elka that one of the killers had gotten away unscathed. Even with the bow, she'd never be able to avenge the deaths and injuries of her people. The man's face had been so oddly painted that she'd never identify him—unless he wore his red tunic. His partner, well that was different. She'd hit him squarely in the chest, and he was either dead, or he'd never be the same again.

As Elka got up to follow Pele, Misha, too, arose looking excited.

"You had a good time yesterday, didn't you?" Elka said, grinning and giving the wolf a pat on the thick fur of her neck. "What are we going to get today?"

Chapter Twenty-Four

"We found some old tracks early on, including some of the wolf's but then..." Palo's shoulders slumped. "...nothing,"

Looking earnestly at his mother and father, Koby followed with, "We checked every path northward. There was just..." Koby shook his head. "There were just no tracks. But we hadn't given up. I mean...we *haven't* given up. We'll find her. But we saw the signal fires and..."

"So, what's up?" Palo asked. "More trouble with the Lion People?"

"No, it's probably worse," Moctu said, frowning.

"The Sea People," Nuri said. "We think they're going to attack us, and they have..." She blew out a breath and shook her head. "Some weapons...some..." Her voice trailed off, and she looked to Moctu.

"Some bows and arrows—strings you pull back, and they send a shaft way faster than an atlatl can. And they're very good with them," Moctu said worriedly. "They rarely miss. We traded for some. I'll show you. They're...amazing."

"So, we needed you back here," Nuri said.

Moctu nodded. "Yeah, we had a meeting of the elders...well, you weren't here, but Ono and I and Nindai met, and we decided to make as many arrows as possible, and some bows if we can. And practice. We'll try and get as good as we can using the bow before the Sea People...before they come. Everyone—men and women—everyone's going to practice, and everyone makes arrows."

"I've been thinking that there's other things we can do," Nuri said. "Other ways to defend ourselves."

Nuri's jaw was set when the men looked her way, and her eyes narrowed as she continued. "Some things that worked when the

Shiv were chasing me." She paused, the thinnest of smiles showing on her face. "We know this area better than they do. Way better. We know every stream, every hill and every tree like the freckles on our arm. We can set pit traps—with sharpened stakes in them. We'll set them along the pathways they'll likely use, and deadfalls and other snares and traps, and we can set up places to ambush them. Knowing this area is our biggest advantage."

"I like that," Moctu said nodding. He had a gleam in his eyes that hadn't been there for days. "I really like that. Whatever you need, even if you have to take people off of making arrows. You're in charge of that Nuri. Great idea."

Late that afternoon, Palo, Koby and Moctu were at the practice field shooting arrows at the target stump.

"I see what you mean," Palo said. "The arrows fly so fast, and I can…I'm hitting the stump or getting close about as often as I can with an atlatl."

"Imagine how good we'll be in a year or two," Koby gushed.

"If we're still here in a year or two," Moctu said quietly. When Palo and Koby looked at him, he forced a quick laugh. "Come on, let's keep practicing."

Chapter Twenty-Five

As Pele's arrow took the antelope in the side near its shoulder, Elka whistled. "That's hard to believe. Wow—what a shot—and from this distance. Are you the best hunter in your tribe?"

Pele was pleased with the shot, and he smiled at the praise. Misha ran ahead to ensure the kill, but the animal was dead when she got there.

Walking toward the downed antelope, Elka was still shaking her head. "Are all the Sea People that good with the bow? That's just..." She blew out a breath.

Pele shrugged, then held up three fingers and wiggled them. He grabbed one signaling that he was among the best. He touched his heart and pointed to the downed animal. "Ver lucky," he said modestly. "Zeke, maybe Arno better." He stared off thoughtfully at the horizon for several moments. "Probably just Zeke."

"I want to get that good. Or close anyway," Elka said seriously, looking up at Pele.

"Take practice, much practice. But..." He held up his arms and flexed his prodigious muscles. "You have strength for it. Most women..." He shook his head. "...can't pull bow this big. And you get better every day."

Elka beamed at his words and hugged his side.

There was a lull as they walked, and the question that had been building inside her for days sprang forth. Even as she was asking it, she regretted her words. "When do you go back to the Sea People?"

He didn't answer but seemed to be considering the matter.

Elka was pretty sure he understood her words, and now the ice was broken. No longer able to keep the questions unasked, she followed with, "When will your people be leaving?"

Pele stopped, and he pulled her to him. After a long hug, he pulled his face away and looked into her eyes. "Much to um…" He waved his hand in small circles, searching for a word.

"Decide?" Elka prompted. "Much to determine? To…to talk about? What?"

"Yes, much to talk about." Pele put his hand to his chest. "Need to trade." He pointed east, and his next words chilled her. "That way…Lion People. Then return my people. Soon."

Elka's heart sank, and she looked at the ground. "Can't go," she said shrugging. "I can't go anywhere near the Lion People. They'd kill me." In a flat voice she continued. "Then you're going back to your people."

Pele put his finger under her small chin and lifted it so her eyes met his. He smiled and again put his hand to his chest. "I want Elka…you. You my mate… Yes?"

Elka's frown crumbled, her dejection snapping the other direction like a willow snare.

"Oh," she exclaimed, covering her mouth with a hand. She dropped her hand and said, "I…I…" but she couldn't find the words. Grinning broadly, she was baffled as to why tears were rolling down her cheeks.

Pele looked at her questioningly as he wiped a tear off her cheek with his thumb.

"No, no…I'm happy," Elka said, hugging him. "Really…I'm so happy."

Chapter Twenty-Six

"THEY'RE COMING…ABOUT FOUR HANDS of them," Zori said, still breathless from a half day of running. "They all have bows, and they look…um…serious, angry."

"Crap," Moctu said scowling. "We're not ready. We need a lot more arrows, more practice, and Nuri says she needs at least two more days to build more pit traps and deadfalls. How far behind you are they?"

"Most of a full day unless they speed up," Zori said. "I ran most of the time, except at night."

"You've done well, Zori. Your father would be proud."

Moctu began pacing, rubbing the beard on his chin and muttering to himself. "Need to slow them down. Talk with them. Can't speak their language."

He stopped and looked at Zori. "Is Zeke with them?"

Zori nodded.

"Yeah, I expect he'd come just to keep them stirred up and hungry for our blood," Moctu said. "And…and he wanted Nuri and some of the other women." Moctu went back to pacing, staring at the ground.

"I'll go," Zori said. "I'll tell them what I saw."

"No, no," Moctu said. "They'd…" He shook his head. "They'd probably just…no, just too dangerous."

"It would slow them down. And I'm willing to try."

Moctu smiled at him. "You're a really brave young man, Zori." He stared at the wall for a moment. "Still, if I went with you, under a fern leaf of truce…maybe we could…" and he was lost in thought again. Several heartbeats later, Moctu nodded to himself then said, "Uh-huh, might work. Only a small chance, but it'll slow them down, and it could save a lot of bloodshed. I suppose they know what waving a fern leaf means. Worth a try. If you really are willing, Zori, we'll go together and try to talk with them."

Zori nodded excitedly, and Moctu continued, "Don't get too excited, Zori. This might just get us both killed. But at the least, maybe we can put some doubt in some of them about Zeke. I'm guessing that some of the Sea People warriors don't like him any more than we do." Moctu patted Zori's back. "You go get some sleep and spend some time with Zaila. We'll leave first thing in the morning."

Moctu was already up and ready to go when the sun crested the eastern hills. Feathery silver clouds edged with pink in a pale blue sky indicated the weather would be mild. There was no breeze, and all was calm and still. Even the birds had yet to start their morning song. The serene beauty of the dawn was soothing, but it couldn't overcome Moctu's troubled mind.

"This is so risky. Am I doing the right thing?" he murmured to himself. "Might get us both killed. Nuri thinks it's a bad idea—too dangerous. Doesn't trust Zeke. I don't either really, but have to try."

Zori set a fast pace, and Moctu struggled to keep up, his body not yet fully healed. It was early afternoon, and they had just reached the top of a hill when Zori pointed and whispered "There they are—some of them anyway. Down by the stream. Some of them have war paint."

As Moctu began waving a large fern leaf high in the air, Zori studied the tree line.

"They haven't split up—they're all there," he said. "And one of them is waving a fern now. Oh…crap, it's Zeke."

Chapter Twenty-Seven

"We go to Lion People. I fix," Pele repeated. He nodded optimistically and repeated, "I fix. You come."

"No. I'm telling you… No, this isn't something you can fix." Elka looked into his dark eyes and saw frustration and disappointment. But she shook her head and in a firm tone said, "No fix." Then she said it again, stressing the words, "*No fix.*"

Since yesterday, when she learned more about Pele's plans, Elka had given the matter a lot of thought, and she continued. "One more thing. The bows…and arrows. You have several to trade. I want those for my people," she said touching her chest. "We trade for those. No bows for the Lion People."

Pele scratched the stubble on his chin. He held up four fingers questioningly. "Your people—four bows?"

Elka nodded confidently, knowing how important the bows would be for her people, and how dangerous they'd be if the Lion People got them. Even after a devastating loss to the Shiv many years back and some costly fights with the Nereans more recently, the Lion People had bounced back strongly, and their population had grown faster and was much bigger than that of the Nereans.

If they had bows…Elka shivered at the thought.

Pele went to his large pack and unwrapped the tall portion that stuck up like a huge rhino horn. Inside, were four, beautifully made bows of slightly varying lengths, all unstrung, each with their sinew cords partially dangling and partially twisted around the individual bows.

"You pick," Pele said, pointing at them. He smiled and said, "I give Elka."

Elka's mouth fell open. Her hand went to her cheek as she looked at Pele and stuttered. "I've never…gotten anything so… Oh Pele, are you sure?

Pele's smile broadened at her delight, and he picked up the shortest of the bows. He gave it to her, saying "This good for you."

With one tip on the ground, it would still come up to Elka's neck, and it was thicker than some of the others.

Pele pointed east toward Lion People territory and said, "Now, you come with Pele, and your people can trade for all bows."

Elka's delight faded, and she shook her head firmly. "No. I'm not going to the Lion People with you. That's something you can't fix. Not going to happen." She extended the bow toward him, giving it back.

Pele looked down at the ground, his lips pursed. But after a moment, his head began to nod in agreement, and he looked at Elka. "You keep bow—I give Elka. You ver strong, not um…" He struggled for the word, and he held his hand about chin level facing down.

Her face flushed, Elka bounced into his arms hugging him tightly. "I'm not tall," she agreed. "This one's perfect. Thank you so much!"

"You keep all bows. One yours, others," He held up three fingers. "Others—I trade with your people—yes?"

"Yes, perfect. Very good," Elka said.

"We hunt now. One more time, then I go," Pele pointed east. "You come part way?"

Elka's joy faded, and the gleam left her eyes. "No, Pele, we're already too close to Lion People territory. It's too risky. If they catch me, they'll kill me."

"I fix," he said, making one last plea.

"No Pele, you can't fix. I'm sorry."

Pele sighed, and paused, frowning for a long moment. "This night, we talk. Maybe I…" He circled his hand around struggling for a word.

"No, I'll be waiting for you at Etseh. I'm sorry I can't stay or come with you. You know I love being with you, and I'll really miss you.

But I can't get any closer… I can't go near the Lion People. There's just no fixing that. The Lion People… They'd kill me."

Pele looked at the ground slowly shaking his head, his lips pressed tightly together. "We go hunt now," he said finally, pointing north and beginning to walk that direction.

To the north, they crested a low hill and broke out of a terrain of tall sedge and shrubs to see a blackened, burned valley. The entire valley was a charred ruin, the devastation running at least to the top of the higher hill far to the north. For a long time, they stared somberly at the sight, which seemed to mimic their mood.

"Lightning fire," Elka guessed. "A bad one."

Pele frowned, surveying the surrounding landscape. "No game north. We go east."

Elka grimaced and eyed him for a moment, to see if he was just trying to get them closer to the Lion People shelter. But they'd come from the west, so hunting a little farther to the east seemed logical, and she nodded her head.

For all the preceding somberness, the hunt went extremely well, at least at first. Pele let out a "Whoop!" when an arrow shot by Elka from her new bow hit the stag square in the neck sending it immediately down to its knees.

"You learn fast," he said smiling and patting her shoulder. "Ver good—you kill."

It was then that it happened. Pele got to the downed deer, and he nudged the stag's antlers with his foot to be sure it was dead. It wasn't. Although still down, the deer lunged his head toward Pele, goring his calf. Elka quickly fired another arrow into it, and the deer breathed its last.

"Barmy, esto barmy," Pele shouted in his language, holding his wounded calf. Elka knelt by him, and he put one hand to his forehead, closed his eyes, and shook his head. "Oh, barmy, barmy."

"Stupid," he said moments later, remembering a word she had taught him. "I stupid, stupid, stupid."

"No, you just…forgot. It happens. The excitement…" Elka said, her voice trailing off. It was a mistake she would never make, especially after hearing how Moctu's birth father, Jona, had died from a similar accident that broke his leg. It was a horrible death—protracted, with high fevers and great pain. Even Ordu, a former leader of the Nereans, had been injured by a seemingly dead auroch. The stories had been told around the campfire so often, that all Nerean hunters were extremely cautious when approaching apparently dead animals.

Pele tried to get to his feet, but decided against it when the pain and bleeding increased. He pointed in the direction of their camp and said, "I fine. You go please. My pack…need yarrow and gallic. Um…" He rubbed his fingers together and then motioned them toward the wound trying to find the word for something else.

"Sphagnum moss," Elka said. "That'll slow the bleeding. I think that's what you want. You rest here. I'll be back soon." With that, she turned and ran, and Misha followed.

It was a sunny day, and Elka was sweating and breathing heavily by the time she reached their camp. She went immediately to Pele's pack and found the yarrow near the top.

Misha crouched beside her, and as Elka dug deeper, the wolf let out a low growl. Elka stopped and looked around, but saw nothing worrying.

"What's up girl?"

Misha quit growling, so Elka resumed her search of the pack. Digging through the supplies of dried meat and some fire-making materials, she found a small skin that held garlic cloves. Elka pushed aside layers of pelts and clothes, some woven, but most of soft, supple hide. Under a pair of old moccasins was another small skin sack, and she pulled it out, noticing it was stained red with ochre. Inside were worn-down blocks of red ochre and white chalk.

"There has to be a medicine kit with moss and willow bark and other stuff," she muttered as she set the ochre aside and dug deeper. "I know he has more than garlic and yarrow." Her arm was buried in the materials of the pack, and she could barely touch the bottom with her fingers.

Elka sighed as she pulled her arm out of the pack, and she began unpacking the upper components so she could better search the lower part of the huge satchel.

"This is taking too long…need to get back to Pele," she said through clenched teeth.

She removed the dried food and a stack of beautifully tanned seal skins. Pulling a folded tunic out, Elka once again heard Misha growl, and she twisted her head to see what the wolf was worried about. Elka knew to take the wolf's warnings seriously. But Misha wasn't looking at an incoming threat, she was looking directly at Elka, and her teeth were showing.

"What's up girl?" Elka said in a lower voice, her leg muscles tightening, readying for action. Holding her breath to better hear, Elka scrutinized the brush and trees around the camp. But she saw nothing, and the sounds were normal, from the leaves gently rustling in the light breeze to the insects enthusiastically thrumming their songs.

"Hmm," she murmured after a long moment, calmer now, but still on edge. "That was weird."

She returned to searching the pack, and that's when she saw it. Elka's head cocked, and her eyes narrowed as she saw the red and white beads of the tunic she was holding.

"What the…?" Her breath caught as the reddish-stained tunic unfolded further, and she saw the brightly colored, red and white beads in a crisscross pattern on the front. Misha was up now, head lowered, snarling loudly with her lips pulled back and her one good eye radiating hostility.

"Spirits, no," Elka said, grimacing as her head jerked backward. Stunned to her core, she found it hard to breathe. The last time she had seen that tunic, the man wearing it was charging at her with a cudgel.

Chapter Twenty-Eight

Moctu continued to wave the fern as he and Zori made their way downhill toward the group of Sea People, all of whom were now standing.

Turn around, turn around *now*, the voice in Moctu's head became ever more insistent. Zeke is still waving his fern. Why is he smiling? He must not know he was seen killing Arno. Will any of his followers believe us? We need to wait to gather the largest group of them together before we tell them what Zori and Indar saw. Maybe some of them will believe us.

Moctu glanced at Zori and saw the young man's face was ashen, his former confidence and courage draining quickly as they approached the well-armed, powerful-looking group. Moctu put a hand to Zori's shoulder and said, "It'll be alright." His lips tight, he whispered. "If you're asked to speak before me, don't tell them what you saw until I give you a thumbs-up signal."

Zori nodded, and his chin lifted as a touch of color returned to his face.

As they drew within several paces, Zeke called to them and pointed to an older man, but Moctu couldn't understand Zeke's words. The man to which Zeke gestured had long, silver-gray hair pulled together and tied at the back. His beard, too, was gray, mottled dark and light around his mouth, but with two long scraggly white wisps dangling from the corners of his chin. His frame was bulkier than Zeke's, but it was mostly muscle. As Moctu scanned the rest of the group, he was disheartened to see how strong and rugged they appeared.

All of them have huge shoulders, Moctu thought. Maybe all the paddling of their boats? Wouldn't want to fight any one of them.

The man Zeke had pointed to stepped forward, and said, "I am Creed-sey-latzu-tok. You call me Creed." He made a half turn and motioned them to follow. "You come sit with us. Safe for now."

Seated around a small fire, Moctu was pleased to see that about half the Sea People band was seated with them, and most, maybe all of the rest were gathered around the periphery. Creed was closer, and Zeke was on the opposite side of the fire. A hush fell over the group, and most eyes turned to Moctu.

Moctu stood and put a fist to his heart. Directing his words to Creed, he said, "Thank you for meeting with us. I am Moctu, leader of the Nereans, and this is Zori, one of our young hunters." After a pause he asked, "Can you understand my words?"

Creed stood, and to the group said words that Moctu didn't understand, except for hearing his name and Zori's. Creed redirected his attention to Moctu. "I understand your words," he said. "My first mate was woman from the Lion People. I learn their language, and yours…uh, almost the same. I translate your words for my people here." He sat back down.

Moctu rubbed the short beard on his chin as he scanned the group surrounding him. Most of them scowled at him with barely contained hostility.

"This could get ugly," Moctu muttered under his breath. He cleared his throat.

In a strong voice he said, "The Nereans, my people, are a peaceful tribe."

As Creed translated, some in the crowd snorted, and there was murmuring and grumbling from many of the onlooking men.

"We think there has been a misunderstanding," Moctu said even louder. "Our scouts have reported your approach—too big a group for a hunting party. And many of you wear war paint and bring cudgels along with your bows and spears. Cudgels are not for hunting, they're for war."

The men of the group stared at him with undisguised contempt, their chins held high, none pretending this was anything other than a war party.

"But you…" Moctu's voice got low and intense as he swept the whole group slowly, pointing to them with his extended hand. "You don't know the whole story…the real story."

After Creed finished his translation, some of the group's insolence was replaced with curiosity, and they were silent, waiting on Moctu's next words.

"Two of our scouts trailed Arno and Zeke as they left us and pulled their travois toward your camp."

After his words were translated, Moctu saw alarm flicker in Zeke's eyes, and Moctu went on quickly.

"Your leader Arno was an honorable man. We liked and respected him. We're aware that he's been killed, and that's likely your reason for wanting war. But it wasn't us that killed him."

As Moctu waited for the translation to be completed, he saw Zeke frowning and biting his lip. The circle became noisy with skeptical shouts. Zeke stood and began waving his arms to forestall any more of Moctu's words.

Moctu pointed to Zori and said loudly, "This young man—and another—saw it all."

Zeke started talking, but Moctu was shouting now as he pointed to the new Sea People leader. "They saw this man—Zeke—club Arno from behind with a large rock. Zeke killed Arno!"

Zeke tried to wave off Creed to keep him from translating, but as the words came out, Zeke turned on Moctu and charged him. At the same time, two younger men, who looked remarkably like Zeke, sprang from the log they were seated upon and hurtled themselves at Moctu, yelling "Non!"

Even though Moctu was expecting Zeke's attack, and he sidestepped and ducked under the man's arms, Zeke managed to snag the collar of Moctu's tunic. Both men were off balance, and they tumbled to the ground, rolling close to the fire. The two younger men were there in a heartbeat, and they fell upon the grappling pair. Zori joined in, pulling one of the younger men to the side and smashing a fist into his nose.

When the other young man's knees landed on his side, Moctu nearly blacked out from the pain and momentarily couldn't breathe.

But he managed to land a punch to Zeke's eye just as Zeke was swinging his fist. Rather than landing a hammer blow to Moctu's face, Zeke's fist glanced off Moctu's forehead. The young man atop him landed a wicked blow to Moctu's cheek, and Zeke rolled from the side so that he, too, was above Moctu. With his left hand, he pushed Moctu's head toward the fire, and he raised his right to strike again, but a strong arm restrained it. Others in the crowd pulled the other men off Moctu and deterred Zori from hitting the man with the bloody nose again.

There was a long moment of commotion and yelling, none of which Moctu understood, but at least the hitting had stopped. Moctu saw Creed, still holding Zeke's wrist, shouting at some of the men who restrained the others.

Moctu winced, and his breath caught as he sat up, and the pain in his ribs exploded. He stayed motionless, his eyes closed, grimacing as he waited for the waves of pain to recede. Zori came to him and from behind, carefully put his hands under Moctu's shoulders.

"Think you can get up?"

When Moctu nodded, Zori helped him struggle to stand. Zeke was yelling at a group of men, and Moctu's eyes fell upon Creed's. His eyebrows were furrowed and his lips pressed together tightly. After a moment, Creed gave a tiny nod to Moctu, then tilted to his head to the side, signaling that it was a good time for them to leave.

"Time for us to go," Moctu said in a low voice to Zori.

Zori nodded, and he quickly gathered their small packs. After a moment's hesitation, he grabbed the wilted fern leaf as well and rejoined Moctu, escorting him up the hill. As they heard Zeke yelling louder, both men unconsciously picked up their speed, but no one followed them.

Chapter Twenty-Nine

"That can't be," Elka said incredulously. "No, no, no, no, no." She put a hand to her face and closed her eyes, but when she reopened them, the tunic was still there. The left sleeve had spots of blood, and it bore the marks of Misha's teeth where she'd savagely fought with the man—the man who was…Pele.

Elka kept shaking her head to rid herself of this horrible nightmare, but it clung to her like pine resin.

"That's why Pele's arm had scratches and scars. And, oh Spirits, that's why the men weren't wearing the typical lion claw amulets. They weren't Lion warriors."

Elka blew out a breath and put a hand to her chest where it ached. Her heartbeat was racing.

"Spirits, I just don't…I just can't believe…"

Memories of their tender lovemaking surfaced, and she murmured, "That was…real." She shook her head. "It had to be. It…*had* to be."

Elka got up and moved to sit on a nearby boulder. Misha came to her and nuzzled her hand.

"Wait, but why would he save our lives?" she murmured as she stroked the wolf's neck. Elka snorted and scoffed as she pushed Misha good-naturedly saying, "You liked him because you were starving and he gave you caribou." She grimaced as another thought hit her.

"That's why he saved me and why he pushed so hard for me to go with him…to go with him to trade with the Lion People. *I* was the most valuable thing he was going to trade. Devu wants me alive so he can sacrifice me." The shock of that thought once more took her breath away, and a tear rolled down her cheek.

"I thought he loved me. I thought..." A sardonic laugh escaped her and she shook her head, wiping a tear from her other eye. She covered her face with her hands and moaned, "Ohhhhhh," fighting the urge to cry. A sob escaped her, then another, and she broke down completely, weeping uncontrollably.

A cold nose on her face, and a rasping tongue licking her tears brought Elka out of it. She held the wolf's head with both hands, and through blurry eyes she looked at Misha's face and smiled.

"He did this," Elka said, her lip curling as she studied Misha's destroyed eye socket. A thought occurred to her, and she knelt again by Pele's pack. She reopened the small, ochre-stained skin and let the contents fall on the ground. There they were—the well-worn blocks of red ochre and white chalk that he'd used as face paint. Their faces that day had been painted with red, white and black. There was no black in the pouch, but they probably just used charcoal.

Remembering that ochre-colored skin had been under a pair of moccasins, Elka rummaged through the pack until she found them. Holding up the right one, she confirmed that it had a hole on the ball of the foot, just like the one she had trailed.

Anger flared in Elka as, for the first time, she realized Pele hadn't just wounded Misha and her, he or his partner had killed the Etseh guard and wounded, and maybe killed, her brother and father. Her eyes narrowed and her teeth clenched tightly as she reflected on that thought.

"He's out there," she muttered. "Wounded. And he's probably wondering what's taking me so long. Maybe he realizes that I found the tunic." She stared off into the distance in his general direction.

"Should I kill him?" She immediately shook off the thought. "Why? Why not? I don't..." She blew out a breath. "Why can't I?" But she shook her head again. Killing him just seemed wrong.

Elka put the side of her fist to her mouth and stared at the ground for a long moment as a plan came to her.

Working quickly, she staked the tunic down in the middle of their camp near the fire. In the middle of it, slightly to the left, just above the red and white crisscrossing beads, Elka placed a larger stake, and with a flat-sided cobble, drove it hard, completely through the tunic.

"That's where the heart would be," she said to Misha who watched her carefully.

Elka gathered up all the bows, and stared at Pele's pack for a moment considering whether to take the finely processed hides. Right now, she wanted to hurt him, to let him know she knew everything.

"My pack's already going to be heavy with the bows, arrows, a sleeping fur, and some of our fresh meat," she muttered, shaking her head. "So, I'll leave him that little stuff to trade with the Lion People. But he won't be giving them the big prize, not going to be giving them me."

With her jaw set, Elka turned and set off to the south for Etseh. The afternoon was cool, bright and sunny, and the leaves on many of the trees had begun to turn gold and red. Larks and finches were singing lustily, and the air was fresh and fragrant, but Elka didn't get far before her thoughts turned gloomy and dark.

At a small pond beside a fast-flowing stream, she knelt to drink and wash her face. The energy Elka had felt as she set off was gone, replaced by heavy feet and a hollow feeling in her chest. She studied her reflection on the still water and was further depressed by her pallid face and drooping shoulders. The burden of sadness was aging her. There were lines on her forehead that weren't there before. Annoyed, she roiled the water to rid herself of the image.

"I should have known," she muttered, tears welling in her eyes. "It was too…" She didn't finish the thought, angrily wiping her eyes with the back of one hand. "I should have known."

Chapter Thirty

"OH, SPIRITS, WHAT'S HAPPENED?" Nuri exclaimed, running toward the two men as Zori helped Moctu limp up the hill toward Etseh. Zaila, Zori's mate, was on the lookout for them, and she'd spotted them as they descended the far hill. Nuri had posted her to be on watch for Moctu and Zori to ensure that the two didn't step into any of her pit traps or trigger any deadfalls. She and much of the rest of the tribe had prepared the traps as defense against the approaching Sea People warriors.

"What happened? Are you alright?" Nuri repeated as Zori disengaged from Moctu to hug his mate.

"We've been so worried," Zaila said, hugging Zori tightly.

Moctu held up a hand to calm them. "We're both fi—"

"They attacked us," Zori interrupted. Moctu was talking, and they jumped us...him mostly. His ribs have swelled up and turned black and blue."

"We've been checking our back trail, and they don't seem to be following us closely, so we may have slowed them a little...given them something to talk about," Moctu said. "We told them that Zeke killed Arno, and that's when all the fighting broke out. But I think some of them believed us." He blew out a big breath. "Ah, but it's good to be back. I missed you," he said as Nuri gently hugged him.

"You're hurt. Let's get you back—"

"Any word from Elka? Any sign...?" Moctu asked, squinting through the forest canopy at a woman digging a hole.

Nuri shook her head and looked at the ground. "No. I'm sorry Moctu." She brought her eyes to his. "I'm really worried about her. I'd go out looking for her if we didn't have this other..." Nuri waved her dirty hands toward the west and scowled. "This other trouble coming. Our scouts are watching not only for the Sea People, but for her, so that she doesn't fall into a trap. If she comes…" Nuri shook her head and started over. "*When* she comes, it'll likely be from the east or north, so she shouldn't have problems from the traps."

Moctu winced as he twisted to look around. "I don't see anything… How's the trap building going?"

Her worries for him eased by Moctu's business-like manner, Nuri moved a step away, and pointed a couple of paces farther along the trail. "That's one…a bad one. I wouldn't want to be the warrior that steps into it. Can't see it, can you?" she asked with a touch of pride. "Avi made that one. It's really going to hurt." She pointed farther up the trail into the trees. "See that deadfall? That log will swing down and hopefully take out a couple of them if they don't see it."

"I can't see it," Moctu said, frowning.

"Good," Nuri said smiling. "That's what we want. Seetu and I worked on that one yesterday morning. Palo's got everyone rotating between building traps and practicing with the bows. He's at the practice field right now." She shrugged. "I've gotten pretty good. But we need more bows and more arrows."

"Maybe we've slowed them enough that we can make more of those. But I think what you've done here…" Moctu pointed up the trail toward the traps, "These traps are even more important. Their warriors are going to be better with the bows than we are, but these…" He whistled. "These traps are our secret advantage. How many traps have you set?"

"There are about four hands of them, mostly on the trails. But if they leave the trails, we've got some in the forest too. And we've cut and spread thorn branches and vines to…to guide them a little," Nuri said, smiling devilishly. She saw Moctu wince again, and gingerly took his arm. "Come on, you're hurt. Let's get you back to Etseh, and let you rest a bit. Stay very close to me so you don't step in a trap." Calling back, she said, "Zaila, you guide Zori along the path, but remember, no leaving tracks—nothing to guide the Sea People warriors. They might even come tonight."

Chapter Thirty-One

"Crap, that only leaves us two," Moctu said after hearing Palo tell him that one of their best bows had cracked after being pulled too far.

"And the three bows we've made don't have the power of the ones we traded for," Palo said. "I'm not sure what we're doing wrong."

"We're trying to learn bow-making in a moon, whereas they've been making them for years," Moctu said, pacing back and forth in the small hearth.

"You need to be resting," Palo said pointing to the sleeping furs. "I'm sorry to bother you with this news, but it's important."

"Can't rest. It hurts no matter if I lie on my side or my back… whatever. I can't even pull a bow now." Moctu blew out a breath.

"You'll get better fast—you're a quick healer. And you need to. You're our best shot," Palo paused. "Except for me, that is." He smiled and winked at Moctu.

Despite his anxiety, Moctu gave a snort of laughter. "We'll see about that when I can pull a bow again."

There was commotion outside and both men exited the hearth to see what was going on.

From the edge of the forest line, Indar was shouting. "They're coming! Two groups. Some along the southwestern trail and some on the western trail. Both of the groups have at least two, maybe three hands of warriors."

Someone shouted an unintelligible question, and Indar answered, "The southwest group just crossed Willow Creek, and the western group just got there."

"Crap," Moctu said again. He looked at the sun which was halfway through its downward glide. They could get here before dark."

Nuri approached quickly her hands still wet from washing the dirt off. "I heard," she said. "We've got traps all along both trails." Smiling she said, "They're in for some painful surprises. Plus, I have an idea that that I think would work pretty well."

Moctu and Palo both nodded, and Moctu motioned her to continue.

"I can take three of our best with bows..." She looked at Moctu and shook her head. "Healthy people only. Palo, we'll need you—and then we can ambush them along the western trail. We'll set up on the knoll where the trail goes along the dry creek bed. Each of us fires an arrow or two then I lead the group back through the traps. When they chase us, the traps will get them."

Moctu closed his eyes and rubbed his forehead. "It *is* a good plan. I'm just not excited about my mate shooting at a big group of expert bowmen."

"Great plan," Palo said enthusiastically. "We've got to do it. Nuri's the one who knows where all the traps are. I helped build some of them, but I don't know where a lot of them are." As Moctu considered the plan, Palo put his hand on Nuri's shoulder and added, "I'll watch over her, Moctu. Won't let anything happen."

Nuri smiled and motioned with her thumb to Palo. "And I won't let anything happen to him."

Moctu chuckled and nodded reluctantly. "Do it."

As Nuri and Palo excitedly hustled out of the hearth, Moctu called after them, "But be careful. Nothing too risky."

Shortly thereafter, Moctu saw Nuri lead Palo, Nabu, and Zori toward the forest line where the western trail started.

Good choice, Moctu thought. Nabu's not only a good shot, he's the fastest person in the tribe. Loves to run. And Zori's a good shot and dependable. Nuri and Palo chose well.

"But they took our two good bows, and two of the three we've made, so that leaves us just one," Moctu muttered. "Hope the group to the south gets stopped by the traps on that trail." He shook his head and blew out a breath. "Otherwise..." He grimaced. "It's going to be us using atlatls against their bowmen. Whew—brutal... It'll be over quickly."

He extended both hands toward the sun, one over the other horizontally with fingers flat, measuring the amount of time until sunset. At full extension, there were two and a half hands from the sun to the horizon. It would take about three fingers of time for Nuri's group to be in position. As he watched his mate and best friend disappear into the forest he whispered, "Spirits, be with them."

While Moctu directed the building of breastworks and barricades at Etseh and the gathering of shafts and atlatls, he twice more checked the horizon with his fingers.

"Well, they're likely there by now," he murmured, nervously pulling on his chin whiskers. "Spirits, please watch over them, and give them good luck. Bring them back safe." Moctu ruefully looked over the pathetic arsenal the remaining Nereans had collected with which to defend Etseh.

"Plenty of darts and arrows, but only one bow, and it's one we made, a not-very-good one. That, and about two hands of atlatls." He bit at a rough edge on his thumbnail.

Moctu had chewed the edge of his thumbnail down near the quick by the time one more finger of time went by. His heartbeat jumped as one of the scouts positioned higher in the hills called out. But it wasn't what he was expecting. It wasn't about the Sea People coming, or about the return of Nuri's group. His head cocked as he was certain he'd misheard. But then the scout yelled it again.

"It's Elka! She's coming from the eastern trail. Elka and her wolf. She's got a big pack."

Moctu's mouth fell open and his heart, already beating fast, began to race. He shook off people attempting to help him, and he broke into a limping run.

"Oh, Spirits, can it be?" He was only now realizing that he'd already begun the process of mourning her. Although he'd never admit it, even to himself, she'd been gone so long that he thought she was likely dead. Rushing ahead, Moctu was wheezing as he crested the hill. Just emerging from the eastern tree line was a person—a woman—and a wolf.

"It's her—it's Elka for sure," he cried, still not completely believing his eyes. "It's Elka and Misha." He began running down the hill to meet her, an awkward limping gait, stumbling twice, but not falling. Moctu felt the wetness on his cheeks, and he wiped away tears, but more streamed down. He'd lost a child, and he thought he'd lost two. Now, he had one of them back. Tears weren't manly, but right now, he didn't care. As Moctu neared Elka, Misha bounded toward him, a man she trusted almost as much as Elka.

"Oh, Spirits. It's so good to see you. Where've you been? We'd just about—" Moctu hit the ground as he was knocked off his feet by Misha's excited onslaught.

Elka squealed with delight at seeing him, and quickly easing her pack to the ground, she rushed to help her father to his feet. He was laughing shakily, but tears still streamed down his cheeks. Moctu winced at the pain in his ribs as the two embraced tightly, happily bouncing a little, with Misha, on her hind legs, trying desperately to join in.

"It's so good to be back," Elka said. "*So* good. I see you've recovered—at least mostly. How's Meko?" She felt her father immediately stop bouncing and his embrace weakened.

"Oh, Spirits, no," Elka whispered as she backed off a step, her hand covering her mouth.

Moctu said nothing. He just looked to the ground and gave a nod, his lips pressed together tightly.

"Oh, no," Elka said, falling to her knees. "No, no, no." Elka's hands covered her eyes, trying to rid herself of the image of Meko's battered body.

Moctu knelt beside her and pulled her to him. He opened his mouth to speak, but no words came out. The two rocked back and forth, and Elka's long moan was gradually replaced by soft sobbing.

After a long moment, the sobbing subsided, and Elka sniffed and asked, "And Mother Nuri—where is she? Is she alright?"

Moctu nodded but looked worried.

"We've got to get back. There's a lot going on—none of it good. We're being attacked by a different group. They're Sea People." Moctu saw Elka's jaw drop, but he was surprised to see her look of shock transform into a knowing and hostile nod. "You know?" he said. "You know of these people?"

"Yes," Elka said through clenched teeth. "Oh, I know a little something about them."

Chapter Thirty-Two

"There's another one over there," Nuri said pointing to a place where the trail narrowed. "It's a really nasty one with sharpened stakes at the bottom. Avi made that one."

The group had positioned themselves on the lip of a lightly forested knoll that overlooked the trail which, for a short stretch, snaked along a dry, sandy creek bed before reentering the woods.

"Palo, when you think it's best, you rise up and shoot, and we'll fire then too," Nuri said. "But don't wait too long. We need to leave fast and be far enough ahead of them so they can't just watch and follow in our footsteps. We'll fire one, maybe two arrows, then break for that thick, white oak there," she said, pointing. "Stay to the right of it, and follow me closely, so you don't hit any of the traps. Watch carefully, and step where I step, but don't leave tracks that they can follow to avoid the traps.

"What if they have a scout, a point man, and he heads up this knoll?" Palo asked.

Nuri frowned. "Yeah, I guess in that case we'll just get to shoot him today before we have to get out of here. But the traps will get some more of them, because…" she blew out a breath and shook her head. "They're going to be angry, and they'll come fast."

Nabu nodded and smiled. "So that we don't waste arrows, if the scout comes up here…" He pointed at Nuri. "Just you and Palo fire at him. That'll keep us from putting four arrows in the guy."

Wide-eyed to this point, Zori gave a snort of nervous laughter. But Nuri and Palo nodded at the suggestion.

Nuri put her finger to her lips as a sound came to her. Someone was approaching. It wasn't long before they saw a lead man step from the far woods and peer carefully around. He was tall, slender, with broad shoulders and a short black beard, and he looked like a young version of Zeke. Nuri's nose wrinkled at the thought of the man who had, some days before, asked Moctu to "use" her. Through the brush at the top of the knoll, she studied "Little Zeke," as she began to think of the lead man.

Probably one of the men who attacked Moctu when he was there with Zori," she muttered softly to herself, her lip curling at the thought. The young man seemed competent and alert, with a bow in his hand with an arrow nocked at the ready. He paused to thoroughly study the terrain before descending the slight embankment into the creek bed.

A sizable group of warriors followed, pausing at the forest boundary, apparently waiting for the lead man to scout the open area along the creek bed. Not waiting for the "all clear" sign, some of the impatient ones began to spill out of the tree cover and move down into the creek bed.

Nuri's heart began to beat faster as Little Zeke began to climb the shoulder of the knoll toward them. From the corner of her eye, on the other side of her, she saw Nabu push Zori's head down after the youth rose up too far to get a look. After a tense moment, Nuri allowed herself an exhale, glad that none of the Sea People had seen him.

By now, Nuri's heart was beating so hard that she was afraid someone would hear it.

"Spirits, please give me strength. Help us with this mission, and get us safely out of here," she prayed softly.

It was time. About half the warrior group had spilled down into the creek bed, and Little Zeke was most of the way up the knoll, almost to where he could see them. Nuri tightened her grip on the nocked arrow in her bow, and she unconsciously nodded at Palo, who's eyes were locked on Little Zeke.

Just when she thought he had waited too long, Palo rose to a kneeling position and fired his arrow at Little Zeke. Nuri followed suit, and both of their arrows struck the scout, Palo's taking him high in the side, spinning him a half turn, and Nuri's arrow striking him in the hip.

Nabu and Zori, likewise, rose and fired at the men below. Nabu's arrow hit a short, thick, powerful-looking man in the leg, but Zori's arrow flew wide. Little Zeke fell backward, mostly out of sight, but not moving. Below, it was pandemonium, with men shouting, assisting the wounded man, and racing for cover while fitting arrows to their bows.

Nuri had just fired her second arrow at the dispersing group as one of their arrows whipped past her head. It was so close she could feel the whoosh of air as it passed by.

"Time to go," Nuri said loudly, not waiting to see if her arrow had hit. Crouching, she turned and headed toward the white oak she had indicated earlier. "Stay low," she shouted as two more arrows whipped overhead. The four of them made it quickly to the white oak, and following Nuri, the group melted into the forest next to the trail.

"How many did we get?" Zori asked. "Pretty sure I got one in the shoulder with my second arrow. My first missed. Nabu hit one for sure." The youth was speaking quickly, almost babbling, coming down from a huge adrenaline surge.

"Everyone alright? Any one hit with one of theirs?" Nuri asked.

"Just a nick," Zori said. "Just a scratch on my shoulder. Doesn't hurt," he said bravely. "Close call though."

"We'll wash that out soon," Nuri said. "But right now, we need to get away."

They moved through the thicker forest, reentering the trail a short stone's throw farther beyond, and they proceeded along it carefully with Nuri guiding the group. They could hear the shouts of the Sea People warriors who had begun their pursuit. They were angry and energized, hungry for blood and retribution.

"Won't be long now—listen," Nuri said, pausing and smiling thinly, anticipating the effect her traps would soon have. Sure enough, there was a cry and a shriek of pain as the first trap was encountered. Nuri nodded and turned to move on, but she stopped and gave a fist pump as another, even louder scream of agony rang out.

"I'll bet that second one was Avi's trap. It was whew…" Nuri blew out a breath. "Anyway, that's two more of their warriors that we don't have to worry about." She looked at Zori. "Now let's get you back to Etseh and clean out your wound. Even a nick can cause problems."

Chapter Thirty-Three

"The Sea People, they have bows that can send shafts way faster and farther than our atlatls," Moctu said as Elka helped him back toward Etseh.

"So, you know about bows," Elka said. "I was going to tell you about them. They're amazing. They—"

"Nuri's out there right now," Moctu interrupted anxiously. "I couldn't go because…" He frowned and looked down, realizing he didn't need to say more.

"Yeah, you need some rest to get healthy again," Elka said.

"I was much better—almost back to normal, then…" Moctu went on to tell her about the truce meeting with the Sea People, which ended with their beating him and rebreaking his ribs.

"Sounds a lot like my dealings with them," Elka muttered.

Their conversation broke off as Elka's grandmother Alta rushed up the hill and embraced Elka in a bear hug, squealing and whooping with delight.

A little farther down the hill Nindai, the lame shaman, smiled and waved at her.

"Oh, we're so glad to see you," Alta gushed. "You had us so worried. Where have you been?" To Moctu she said, "Indar just came back with a scouting report on the southwestern trail. That warrior group has passed most of the traps we set. But he heard a lot of screams—says we hurt at least three of them, and slowed them way down."

Moctu nodded and looked to the sun, noting that it would set shortly. "I think we've got until morning before they come. With all those traps, they won't risk moving along that trail in the dark. Nuri's group ought to be back soon with their bows. We'll need everyone to have a chance.

"I have bows," Elka offered.

Moctu's eyes widened. "You have a bow? Wait, you have *bows*?"

Elka nodded and smiled at his excitement. She held up four fingers. "I have four of them."

Moctu studied the large pack she carried with the tall, skin-covered part that stuck out. "Where…how in the…" He shook his head and laughed. "Never mind, we'll get to that later. Come on, let's hurry."

With that, he hobbled faster down the slope shouting, "More bows—we have more bows. Elka brought bows!"

As they approached Etseh, Elka saw that the shelter was humming with activity. Women collected water in skins while other people rolled or carried logs up the hill for further fortifications. Elsewhere, women held arrows over fires while using batons with holes at the end to bend them, helping to straighten the shafts. The shelter was now as fortified as she had ever seen it. Large rocks and logs had been stacked, and poles with skins completed a wall that blocked all of the easy approaches. As for the steeper approaches, large logs were positioned and ready to release to roll down the slope. Behind a low wall was a huge pile of stones that were just the right size for throwing.

Moctu beamed as Elka unwrapped the four, fine bows. He yearned to keep the longest one for himself. It was a beautiful cream color, very smooth with well-shaped curves. But his rebroken ribs would not allow him to pull the bowstring more than halfway. Elka kept her shorter, thicker bow, and the rest they parceled out to those in the tribe who had practiced the most. Moctu gave out his cream-colored favorite last, telling Indar that he was just loaning it to him for the time being.

A cheer went up as Nuri's group emerged from the west forest and made their way back to the fortified camp. It wasn't long before the foursome was back at the Etseh shelter.

Nuri immediately hugged Elka, and the two rocked back and forth whispering as tears rolled down Nuri's cheek. Seeing Moctu's inquiring look, she reluctantly released Elka and turned to face him. "We got four of them at least. Probably three just wounded, but

Little Zeke, uh, I mean what looked like a younger version of Zeke, well, my guess is he's dead or going to be."

Moctu nodded. "Sounds like one of the ones that jumped me." He thought for a moment, then smiled. "Tabar used to say that wounding an enemy was like wounding two, because someone had to take care of the wounded one." He paused, scratching his short beard, doing the mental math in his head. "Indar heard at least three screams on the southern trail, so that's probably another handful of them out of action. They probably left a couple to guard their boats and camp, so they're down by at least half their number." Moctu hugged Nuri tightly and for the first time in days felt some real hope. "Your traps, the traps and your ambush may have saved us, Nuri. You're…" He shook his head. "Have I ever told you that you're amazing?"

Nuri smiled broadly and blushed a little. "No, never," she lied.

"But adding those numbers, so many of them out of action, that's given me an idea." Moctu's eyes narrowed, and he looked off to the side. He nodded slowly, and a sly smile crept to his face.

Chapter Thirty-Four

"The odds are much better thanks to Nuri's traps and the ambush," Moctu said smiling at his mate. He, Nuri, Palo and Nabu were sitting in his hearth, across from glowing embers in the stone-encased firepit. "We're now about evenly matched in numbers, and thanks to Elka, in bows too."

"So now we just wait for them to come?" Palo asked, frowning.

"No, first off, I'm worried about them trying to come at us from behind—from the east where we have no traps. So, I've sent Leuna's son, Itzal, to the high, eastern lookout until tomorrow afternoon to watch for that. He's young, but he can handle that."

Nuri nodded vigorously. "I've been worried about that too."

Looking at Palo, Moctu said, "And we're not going to wait for them to come, we're going to attack." Moctu grinned at the surprise on all their faces. "We know this area so much better than they do. Palo and Nabu, you're the fastest. I want you to choose one more warrior and get around behind them. Here's my plan." As Moctu elaborated on the details, their looks of doubt transformed into knowing nods to each other.

"Just might work," Palo said, his eyes twinkling. "Just might work."

That afternoon, Palo, Nabu and Zori gathered supplies and melted into the forest taking a rarely used route that Nuri assured them had no traps.

Dark clouds rolled in and the sun was halfway down when the Sea People warriors showed up on the southern route. Shortly thereafter, the group on the western trail arrived. There were more of them than Moctu expected, and he bit his lip as both groups edged closer, still out of bow shot.

"Hmm, going to have to stall them," he murmured to himself.

It was twilight when someone began shouting from the southern group, and Moctu went to the barricade on that side to hear what was being said. Nuri was there, and said, "They've moved a little closer. Zeke's saying something in his language."

Moctu watched Zeke speak to the man he knew as Creed, and Creed yelled, "Zeke says just give us three of your women, and we will leave you in peace. No more need to get hurt."

Zeke said a few more words to Creed, and the translator shook his head no. Zeke became more insistent, and Creed cleared his throat loudly and reluctantly said, "Zeke says Nuri is to be one of the women."

Nuri immediately looked to Moctu and saw his eyes narrow and his lip curl.

"Come try and take them, and you and your men will soon be feasted on by flies and worms," Moctu snarled back loudly. "We're going to let your bodies rot where they fall."

Creed translated for Zeke, and in the darkening light, Moctu saw the Sea People leader nod to a huge, beardless warrior who carried a long, thick bow. The powerful man drew back an arrow and released it. Almost instantly, with a loud 'thunk', the arrow embedded itself in a log of the barricade directly below Moctu.

"Will they come now?" Nuri asked. Moctu heard no fear in her words, just curiosity and maybe even excitement.

"I don't think so. Their advantage is their bowmen, and they need light to see what they're shooting at."

"I have another idea," Nuri continued. "I'm a Nightwalker—you know I can prowl silently in the dark. And I know every bush and stone and tree around here as well as I know my own fingers. Let me sneak down there and slit one of their guard's throats. That'll have them trembling in the morning."

Moctu smiled at the thought and at his mate's fearlessness. But he shook his head. "I like that thought, but it's too dangerous. Not going to risk you. And you're probably the thing, the woman that they want the most. At least Zeke does. So no, please don't do that. And I know you—when you get an idea like this. Please give me your promise not to do that."

"Alright, I promise," Nuri said frowning a little, but hugging him at the same time.

Chapter Thirty-Five

The three men were moving quietly through the forest, and they were well behind the Sea People warriors when they heard talking. Nabu alertly stopped, and Palo put a hand to Zori's chest, slowing him, and a finger to his lips signaling absolute silence.

The noise had come from their left, and as they waited to determine what it was, Palo lowered his pack and soundlessly strung his bow. The two other men followed suit.

It was two men—warriors of the Sea People—and one was helping another who limped with a bad right leg.

"Ow!" said the injured man, and he followed up with a loud string of what had to be curses.

Zori tapped Palo on the shoulder and moved closer indicating he had something important to tell Palo. Palo frowned but bent down, and Zori whispered directly in his ear, "The wounded guy is one of the ones who attacked Moctu. Think he's one of Zeke's sons."

Palo scowled and nodded. He placed an arrow on the string and began to draw the bow, but Zori shook his arm and hissed, "You're just going to kill them, an injured man?"

Worried they'd be heard, Palo shushed the young man and angrily shook his arm away. In a lower voice he hissed back, "Yes, I am—both of them."

Nabu had an arrow on the string, and he nodded to Palo that he was ready, but he looked alarmed at the noise Zori was making.

Once again, Zori slowed Palo, placing a hand on Palo's bow. "Shouldn't we just take them prisoner?"

Palo shook him off again, but it was too late. The two Sea People warriors had heard them, and an arrow whizzed past Palo, taking Zori in the upper arm and spinning him to the side. Nabu drew and let loose an arrow taking the uninjured warrior in the groin. The man went to his knees clutching the arrow, then fell to his side.

Hampered by his bad foot, the injured man was drawing his bow when Palo's arrow hit him in the side of the neck, passing most of the way through. The man was thrown backward, and he released his arrow upward, harmlessly into the high forest canopy.

While Nabu moved forward and dispatched the two down men, Palo turned his attention to Zori. He was as angry at the young man as he was sorry for him.

"Oh crap, Zori, what have you done? You've put the whole mission at risk. But let's take care of that arrow in you…before I smack you."

The arrow was halfway through the meat of Zori's upper arm. Palo felt the arm and said happily, "Didn't hit the bone—that's good. I've dealt with this before. In fact, it happened to me years back. Bite the sleeve of your other arm—this is going to hurt."

In moments, Palo had broken the arrow in half and extricated the piece with the tip from one side and the feathered portion from the other. After watching the wound bleed, he smiled. "It's bleeding a lot, but the big vein wasn't cut. That's good. Just enough bleeding to clean most of any poison or dung they put on their arrow, but not enough bleeding to kill you."

Nabu came back and nodded to Palo. "Took care of them. Pulled their bodies off the trail into the forest. We have two more bows and two hands of arrows."

Palo nodded, then tore a strip from his tunic, and after wrapping Zori's wound tightly, he looked sternly at the young man.

"Nabu and I are going to continue without you. You're going to walk back to Etseh, but you're going to have to be careful of two things. The Sea People are there by now for sure, and they'll have warriors surrounding it in various places. You're going to have to move slowly and silently, especially when you get close. You may be lightheaded or confused by the time you get there from loss of blood or if there's poison."

He considered the two new bows. "You'll take your bow plus these two with you. They'll need them at Etseh. Now the other thing you need to be careful of is this—our people are going to be tense and jumpy. There'll be a point where you could be shot by either group, them or us. But you know the land around there—you know every ravine, every tree, every bush. Use that knowledge. Just get there safe, and clean that wound as soon as possible."

He went to Zori's pack, and he pulled the water skin and some dried caribou from it. "You take the water and meat. We'll be taking your pack. Need a bunch of those supplies." He patted Zori on his uninjured arm. "Now get going. Nabu and I need to move on."

Zori groaned as he got to his feet, but he took the bows and his meager supplies and began walking back the way they'd come. The two older men watched him for a few moments, then looked at each other, shaking their heads.

"Can we do this—just the two of us?" Palo asked.

"Have to," Nabu said simply.

Chapter Thirty-Six

Itzal returned at midday, earlier than expected from his lookout duty. Tall for his age, the boy had stripped off his tunic even though it was cold and breezy. He was dripping sweat and breathing hard as he rushed to where Moctu manned the barricade.

"They're coming," he said loudly, pausing to take another deep breath.

Moctu's head cocked, and he stared in disbelief at the youth whose head was held high, proud to be delivering such important, if troubling, news.

"The Sea People warriors are behind—?" Moctu started.

"No—no—the Lion People. It's the Lion People. Looks like two, maybe three hands of warriors…coming this way. They're in the lowlands, oh, about two days away, probably more."

Nuri, Elka, Lehoy, and others crowded around to hear more about this worrisome development.

Moctu looked away and upward, shaking his head.

"Spirits," he said. Then, to himself as much as to anyone nearby he wondered aloud, "Could it be a hunting party?"

Rubbing his forehead, he muttered, "Doesn't seem likely." With a frown, he forced a laugh and said, "Just what we need."

Turning to the youth, he said, "Itzal, you did well. For the moment, stay here. We may need you at the barricade. Tomorrow, you'll need to go back to the eastern lookout to check their progress."

Heading back into his hearth he murmured, "Now, I need to think."

Nuri followed, and she found him seated on a log staring down at the dirt floor. He looked at her, and his hands turned up.

"We should be out hunting. Our food won't last another moon, and winter's coming. Instead, we're sitting here between two forces that want to kill us. And each of those forces have more warriors than we do." He pursed his lips, then said, "And I sent three of our best warriors away. We're going to need Palo, Nabu and Zori here over these next few days."

Nuri smiled reassuringly. "They're out there, and they're going to cause the Sea People some real problems. I still like it, Moctu—it was a good plan. And we're as ready as we can be here at Etseh. All we can do is ask the Spirits to be with us."

Later that night, Nuri couldn't sleep, and she left her furs to walk the barricade.

"How are you two doing out here?" Nuri said as she approached Elka and her wolf Misha. It was pitch black as the cloud cover blocked out any light from the moon and the stars. Only a few of the nearby hearths had fires going, and their entrance skins were pulled closed. Nuri could barely see the pair, but could make out their movement.

Elka had offered to patrol the shelter with Misha during the night to warn of any attack and ward off any small bands of warriors that might test the barricades.

"Doing fine—my stomach hurts from all the tension. Misha's curious what we're up to and why we're not sleeping."

Nuri bent to stroke Misha. "I can't seem to sleep. Want some company?"

They walked in silence until they reached the southern barricade, and when they turned, Elka said, "I heard two of the women talking today. They think I've caused this…this situation with the Sea People. And now the Lion People are coming too. Everything's my fault. Maybe that's true."

Nuri put her hand on Elka's shoulder. "No, that's not true. If anything, you've saved us. You brought us four bows, and that's what we needed most. Don't let yourself think like that Elka."

"It's hard…it's hard not to think… Crap, even I think they're coming because of me."

Oh Spirits, she thought to herself, feeling a little dizzy. What if they're being led and helped by Pele?

They walked in silence until they got to the northern end, and Nuri said, "You were gone a long time, and you haven't talked about it with anyone. Misha lost her eye, and you came back changed too. You for the better, I think. And the bows. Will you tell me a little of what happened out there?"

Turning to walk the other direction, they were halfway to the southern barricade before Elka spoke. "I killed a man. At least I think I killed him. One of the two that attacked us. The other man…well…" Elka paused for a long moment wondering what to say about Pele. There was just so much. The man who'd been involved in killing her brother, Ender the guard, and who'd almost killed her father as well as her and Misha. But also, the man who'd saved the pair of them from certain death, who was her first kiss, and the only man she'd ever loved—in that way. But then again, the one who'd deceived her and tried to get her to go with him so he could trade her to Devu and the Lion People.

"Well, he's a weasel—a wolf—a mean one, and he got away," she finished.

"But you got the bows from them?"

"Yeah."

"Well, you avenged Meko and Moctu, and you brought back the bows that are our best chance against the Sea People—and the Lion People, if they come. You're a hero Elka."

Suddenly swamped by guilt, Elka felt her stomach turn, and she bent to throw up, but nothing came. She coughed, then stood and began walking again.

"There are other things," Elka said in a voice barely above a whisper. "Things I want to tell you, Mother Nuri but…" In the darkness, she shook her head. "But not now. Maybe another day."

Nuri patted her back. "Any time you want my dear. Any time."

A cold breeze began to blow hard from the north, and Elka realized she could faintly see the far tree line to the south. The gray light of dawn was coming. And there was movement in the trees.

Chapter Thirty-Seven

ELKA STARTED IT, BUT the cry soon echoed throughout Etseh.

"They're coming!"

As Sea People warriors poured out of the forest from both the southern and western pathways, sleepy Nereans assumed their places at the barricade. Moctu had slept fitfully, but now he felt totally energized. His heart was beating hard, but it was as much from excitement as fear. The odds were still against his Nereans. The Sea People were just so much better with bows. But every other factor was stacked in favor of Moctu's people. They knew the terrain better. They could fire their arrows and atlatl darts from behind cover, whereas the Sea People were mostly exposed. Even the sun had now crested the eastern hills, and it would be in the enemy warriors' eyes. The Nereans had the higher ground, and they had some surprises ready for the Sea People if they got close.

By Moctu's count, at least eight warriors had emerged from the southern approach and nine from the western one. All carried bows that were strung and had an arrow in place. The attackers stopped just out of range of Nerean bow shot.

"Last chance," Zeke yelled, then he said some words to Creed and gestured for Creed to convey them upslope to the Nereans.

"Zeke says we can make it easier on you," Creed yelled. "Just give us two of your women, and we will leave you in peace."

Zeke said some quick words to Creed, but Creed frowned and shook his head no. Zeke stepped menacingly closer to him, and Creed complied, shouting, "One of the women must be Nuri."

People looked to Moctu, and he stood to reply, but to his left, Nuri rose above the barricade and yelled, "That's not going to happen."

Creed relayed Nuri's comment to Zeke who nodded as if he expected that answer, then he motioned his warriors forward.

At the sight of nearly twenty heavily armed, powerful-looking, expert bowmen making their way up the gentle incline, Moctu swallowed hard and whispered, "Spirits please give me strength and protect my people." He heard people all along the barricade murmuring similar prayers.

"Let them get closer," Moctu shouted. "Wait 'til you can see their war paint. Make your arrows count."

Once within bow shot, the well-disciplined Sea People warriors began rushing forward in zigzag patterns, occasionally launching an arrow at any head that popped over the barrier.

"Now!" Moctu yelled, and seven arrows and several atlatl darts were sent downhill. All but one missed, most going high. Elka's arrow took a dark, heavily bearded warrior in the leg. She watched him cry out and go down rolling to his side, clutching the arrow in agony. She quickly ducked back behind the barricade, eyes wide after an arrow whipped by her head.

Indar was not as fortunate. After making his shot, two enemy arrows came his way, the second thudding sickeningly into his left shoulder and knocking him backward onto the floor of the overhang. In moments, his mother Avi was there tending him, tears already streaking down her cheeks.

In short order, Moctu's stepfather, Alfer was deeply grazed in the right shoulder putting him temporarily out of action.

And elsewhere, an errant shot over the barricade hit Leuna in her bottom, just after she'd turned to get another load of arrows.

With the warriors within an easy stone's throw, Moctu worried that Etseh was about to be overrun. He grabbed cream-white bow Indar had been using, and even though he couldn't manage a complete pull, he fired several arrows. One of his, and an arrow from the lame shaman, Nindai struck two more enemy warriors. Neither hit was lethal or likely to put the men out of action, but both warriors turned and fled downhill.

Alta, Ono and some of the other older people of the tribe began throwing rocks downhill, and one of them hit a third warrior who also turned and retreated. The man next to him put an arm up to help dodge two more of the thrown missiles. He backed a few steps then turned and followed his friend. Seeing a number of their companions retreating, all the other Sea People warriors did likewise. Zeke was last, pausing to scowl and shake his head before he, too, zigzagged downhill.

"Whew, that was close," Moctu muttered to himself. "Too close." Then he shouted, "Who's hurt?"

"Indar's down. Nindai… Jelli we need you!" Nuri called for the two healers as she knelt by Avi who sobbed as she cradled her son's head against her very pregnant belly on what remained of her lap. The thick arrow still protruded obscenely from the young man's shoulder. The rhythmic movement of his chest indicated the motionless boy was passed out, not dead. Nuri gently pulled on the arrow with no success.

"I've tried that," Avi said, her voice quavering. "It's…it's hit the bone. Won't come out."

Nindai and Jelli, you're needed here, too," Alta called loudly. "And Alfer's down. Arrow sliced his shoulder."

"Please, come help," Seetu yelled while tending his mate. "Here too. Leuna's badly hurt."

Moctu grimaced and shook his head. "Two of our warriors down and a woman badly wounded. And we only hit only one of them seriously from what I saw. Don't have many warriors left." He began doing the count on his fingers. "Nuri and Elka shot well, and Nindai did too. But he's going to be busy tending the wounded. Lehoy's okay, and Seetu—but his mate's hurt. And I can barely pull a bow."

Moctu squeezed his forehead hard, trying to ease the tension. "Lion warriors on the way. We need Palo, Nabu and Zori—shouldn't have sent them off." He looked to the western horizon where dark, mottled clouds mimicked his mood.

"Need time. Got to have more time."

Chapter Thirty-Eight

It was a warm day for late Fall, and the sky was gray with darkening clouds. Intermittent lightning flashes between the clouds indicated it would be storming soon.

"How many do you count?" Palo whispered to Nabu.

"I just see two of them, and the three women."

"That's all I see too. Seems like there should be more," Palo said frowning.

"Do we wait and watch some more?" Nabu asked.

"If we wait too long, more will be coming this way bringing back the ones that got injured in Nuri's traps. I think we gotta—"

"Look—movement over on that hill to the north," Nabu said pointing.

"Yeah, that's another one," Palo said, his frown deepening. "Got to take him out first. And quietly."

Nabu nodded, and they immediately began moving that direction. By the time they were in position, scattered drops of rain began to fall—just a few, but they were huge, and increasing in number.

"Rain's going to help," Nabu whispered, smiling as the big drops began drumming into the forest canopy.

Palo nodded, and holding a large rock, he pointed to himself then to the guard, who was sitting and leaning against a large beech tree about six paces away. Next, he touched Nabu's bow and motioned with his head, indicating Nabu should shoot if he had, too.

Any noise Palo made as he crept toward the guard was drown out by the beat of raindrops on the beech tree's broad leaves. Additionally, the guard had covered his head and shoulders with a small skin to shield himself from the dripping of water through the leaves. Palo did well, and it was over quickly.

"May have hit him a little too hard," Palo said gesturing to the prostrate guard as Nabu walked up. "Let's tie him up anyway."

Rain began to pelt down harder and faster, but that ended up making the job easier for Palo and Nabu. The three women and two remaining guards sheltered under a large group of skins that were draped over three boats that were stacked lengthwise like tent poles.

Their bows drawn, Nabu and Palo motioned the guards and women to lie on the ground, and then they tied them tightly.

From the camp, even through the downpour, they could see the other boats, all well up on shore above the debris line of high tide. The boats were well-secured, either on their side or upside down, a pair of them tied with rope to each of seven firmly planted stakes. After a quick inspection of the camp, Palo and Nabu decided to take only the bows they could find, of which there were three finished and two under construction. They pushed the lean to over and dragged two of the three boats down near the others. The third had two large holes in the bottom, and they left it.

They untied all the boats, then grouped seven behind each of two of the most seaworthy-looking craft. Several boats had skins covering the compartments where the seats were, and Palo and Nabu found similar skins they could use to weatherproof some of the other boats.

The rain lightened up, and using shredded cattail and a little kindling from their packs, Nabu built a fire under the disabled craft back at the camp. He ignored the unintelligible cries and outrage of the bound captives, as he piled wet branches and leafy boughs over the boat. Although all the wood he gathered was damp, the inside frame of the boat was dry, and the pine resin and bitumen coating on the skins caught easily and burned well. The fire was soon putting out huge clouds of dark, gray-brown smoke. The storm was moving southeastward, and Nabu stared in that direction for a long moment. The skies were murky and darker toward Etseh, and he worried that even though the column of smoke was enormous, it might not be visible to the Nereans.

Can't be helped he thought. It'll certainly be seen by Sea People warriors returning with their wounded. We'll build the two signal fires when we get the other boats to safety. With that thought, he hurried back to the shoreline.

The rain now was intermittent, and sunlight broke through spaces in the dark clouds, gilding their edges with bright, white gold. Large, frothy, dark waves continued to roll into the shore, but now the sun's rays caused scattered areas to glisten dazzlingly.

All the boats were large—each easily capable of holding two individuals, and grouped together, they were a formidable, perhaps impossible challenge. Nabu and Palo grimaced at the huge cluster of boats, and the thought of what they were intending to do.

Looking at one another, Palo shook his head uncertainly. Nabu shrugged.

"Can we do this?" Palo asked, his voice an octave higher than normal.

Nabu shrugged again. "Have to."

"At least we don't have far to go, and we'll be headed southward with the current," Palo said, trying to reassure himself. "I'll help you get your group out past the surf, then I'll try to get mine."

Looking dubiously one more time at the boats and surf, Palo began shoving Nabu's group into the cold water, and his laconic friend joined in.

Nabu was soon paddling his boat guiding his group of empty boats out past the surf.

Moving back toward his own batch of boats, Palo blew out a breath. "Now comes the hard part," he said, chuckling nervously while pushing his lead boat into the surf. "I've never been in a boat."

Chapter Thirty-Nine

"The rain probably slowed them down," Nuri said to Moctu. "Their clothes, arrows, bowstrings…everything of theirs is wet. Probably slowed the Lion warriors too."

Moctu nodded and looked upward at the dissipating clouds. "Thank you Spirits for slowing them." Searching the western skies, he saw nothing in the few breaks between clouds. He looked back at Nuri and said, "Yeah, that's helped, but we still need more time. How are Leuna and Indar doing?"

"Leuna's way better off than Indar. Besides hitting the bone, his arrow seems to have had more poison on it. Avi's really worried about him. High fever. As for the other one…" Nuri eyes narrowed watching Moctu, knowing he didn't think much of his stepfather. "Alfer's going to be fine, but won't be able to pull a bow for a while."

"There they are," Moctu said loudly, pointing at warriors emerging from the southern forest.

Moments later, another group assembled from the western woods. Moctu's quick count indicated that together, there were three hands of them. Yesterday, there had been three hands plus two.

"That's progress, I guess," Moctu said to himself. He was lost in thought for a moment, looking at the ground and scratching his beard. He turned his attention to Zeke's group as the Sea People Leader yelled some foreign, incomprehensible words at them. Moctu saw Zeke turn to Creed for the man to translate his comments.

Moctu hurriedly grabbed Nuri's arm.

"We have to have more time. Will you trust me as I say some things to them that you might find…uh, disagreeable?"

Nuri wrinkled her nose, but nodded slowly.

Creed yelled, "There's no reason for any more of your people to be hurt. Give Zeke…um, give us what we want, and we leave you in peace."

"Tell Zeke his words are as meaningless to us as a boar's fart," Moctu yelled back, prompting laughter among the Nereans. "If he had any balls, he'd fight me one on one for the right to Nuri. Just man-to-man—no weapons. But he's not enough of a man. Last time he needed his friends and sons to beat me. And earlier, he had to sneak behind your leader, Arno, to club him with a rock.

Creed blanched but turned to Zeke to translate. Moments later, Zeke began charging up the hill before two of his men restrained him. Red-faced and still wrestling with his men, Zeke yelled at Creed and motioned with his hand toward Moctu.

Nuri took Moctu's arm and whispered, "You can't fight him—you're still not healed."

Moctu gently shook her off and said, "Trust me, just trust me."

Creed yelled back, "Zeke says anywhere, anytime, anyplace."

Still well out of bowshot, Moctu stood on his barricade and yelled, "Tell Zeke I need one more day to heal. And this time, he won't have three of his men to help him. Tomorrow…when the shadows begin to lengthen to the east."

After translation, Zeke sneered and held a fist up toward Moctu for several heartbeats. Then he and his men turned and reentered the forest.

As Moctu got back down off the barricade, Elka and Alta were there with Nuri.

"You can't fight him—you're still limping, and your ribs are broken."

"Don't do this. He's healthy, you're not."

Moctu smiled and extended his hands, palms outward. "I can beat him, don't worry. But I don't think I'll have to. Just wait."

Chapter Forty

Night had fallen when Zori made it back.

"Don't shoot—it's Zori," he had called over and over as he came up the hill within bowshot from Etseh.

"I'm wounded, but I have three bows," Zori said with a touch of pride.

Moctu frowned, realizing now why he hadn't seen smoke on the horizon. Their mission had failed.

"And what of Palo and Nabu?" Moctu asked, fearing the worst, but confused by Zori's upbeat tone.

"Oh, they went on," Zori said, and he related what had happened.

As a sleepy Nindai treated Zori's wound, Moctu began pacing the barricade again.

"I don't think two men could do it," he murmured, rubbing his beard. "That's why there's been no smoke." Moctu stared into the night thinking of his two friends. "Spirits—I hope they're alright." Then he shook his head, and through clenched teeth he said, "Crap, crap, crap."

In the darkness, thinking of tomorrow's fight, Moctu gingerly pressed the ribs on his left side. "Gonna have to do this, and oh, crap—its' going to be brutal," he murmured. "Palo, Nabu—Spirits be with you. Please—if anybody can do it, you two can. *Come on.*"

As dawn broke, and the larks, sparrows and robins began their morning songs, Moctu was still up pacing. To the west, he saw scattered white and pink clouds in a pale-blue sky. But there were no columns of smoke. Unconsciously touching one of his ribs, he frowned and pursed his lips.

"Looks like I'm going to have to do it," he muttered to himself. "Can't back down." Shaking his head, he repeated the word he'd said all through the night as he tested his ribs. "*Crap*."

"You were up all night?" Nuri asked, approaching from behind. "Still no smoke?"

Moctu shook his head.

"Moctu, you can't do this. You're not yourself, You're—"

"Gotta do it," he said shrugging.

"Well, ask him for another day or two. Give yourself a chance. You know, besides caring for you, I have a stake in this too."

"I know. I didn't think I'd have to do this, but I won't…I won't let you down. I can win this. I *will* win this."

"I'll just tell you—I'll never be with that man. And if somehow, it works out that I must…" Nuri paused, and her eyes became steely. "Well, he's going to end up with his throat cut."

Moctu smiled reassuringly and hugged her. "It won't come to that." But even as he said the words, he checked the western sky for smoke.

The morning wore on and the western sky turned a beautiful azure blue, dotted with small white clouds. Pretty, Moctu thought, but it'd be a lot prettier with two trails of gray smoke.

Moctu's pulse quickened as he saw Sea People warriors begin to emerge from the forest and assemble at a broad meadow about midway from the tree line and Etseh. Zeke wasn't with them yet, and Moctu rolled his eyes, suspecting that Zeke wasn't saving his energy, he just wanted to make a grandiose appearance.

Even though Elka, Lehoy and others had gathered to walk down to the meadow with him, Moctu ruled that out.

"The barricade on this hill is our biggest advantage. I don't trust them—they could start shooting if you're down there, and it would be over quickly."

As it registered that he was really going to have to fight today, Moctu began recalling everything he'd learned from wrestling matches with the Krog. The Pale Ones fought the boredom of long, snowed-in winters by occasionally having wrestling matches. At their insistence, Moctu had participated, even though any one of them could likely have killed him quickly. They had restrained themselves when fighting him—for the most part—and he'd learned a lot.

Nuri and Elka, then later Alta and Zaila all visited Moctu and implored him not to fight Zeke. He listened and tried to calm them, but he was adamant that he had to fight.

A quick look at the sun indicated it was time. Moctu studied the horizon one more time, then stripped off his belt that held his knife. As he climbed over the barricade, he heard his mother Alta sobbing.

"Not the triumphant sendoff I was hoping for," he mused. But thereafter, a few cheers went up from the Nereans, then more joined in, and they continued while he made his way toward the gathering of Sea People warriors.

As he suspected, Zeke appeared just as he got there, and the warriors cheered and beat their spears on the ground. Despite the cold breeze, Zeke was naked from the waist up, and his broad shoulders and sculpted physique were impressive. He sneered, lifting his chin as he saw Moctu look his way.

Moctu stripped off his tunic to match Zeke. The two men each had scars that indicated they'd fought tough battles, both with other men and fierce animals. In addition to a nasty scar on his left shoulder, where he'd long ago been hit by an atlatl dart, Moctu still had dark, bluish-gray bruises on his ribs.

In his foreign language, Creed made loud and stern comments to Zeke and the semi-circle of warriors. They listened, and when he finished all, including Zeke, nodded in agreement. Then he turned to Moctu and said, "Here are rules. Fight to death. No weapons, just fists and strength. Zeke wins, Nuri and one more woman come to us, then we leave your people—all peace and friends. You win, we leave—all peace and friends. Anything else is dishonor. Big dishonor. You understand and agree?"

Moctu nodded, wondering what Nuri would do if he lost. Not going to lose, he thought, gritting his teeth. As soon as he nodded, Zeke charged him, slamming his hands into Moctu's chest and bowling him over backward while the surrounding warriors cheered. Moctu landed with a thud, and waves of pain rippled through his ribs, but he bounced up quickly. Zeke was immediately there and swung a fist at his face which Moctu ducked. Even so, the fist grazed off the top of his head, stunning him.

The two men circled each other for several moments, and Moctu studied the brawny man, desperately searching for a weakness. Head down, Zeke rushed him again, but Moctu dodged the charge and pushed him off balance. It wasn't enough to make Zeke fall, but it gave Moctu a clue on how to fight this bull of a man.

The larger man switched tactics balling his fists and stalking Moctu as he tried to stay out of reach. Zeke sneered and started saying things that the surrounding warriors began to echo. Zeke was in mid-sentence when Moctu saw an opening, and he waded in, slamming a fist into Zeke's nose and another meant for Zeke's chin, but which hit the big man's eye as he ducked. Zeke responded in kind, his left fist hitting Moctu's ear and his right hitting Moctu's damaged left ribs.

It was hard to tell which hurt more. Moctu's ear was ringing, but his ribs were on fire.

Can't take many of those, Moctu realized, as the two men began to circle again.

Zeke said something to his men and they began chanting, "Stat tu, stat tu!" which Moctu took to mean 'Kill him' or 'Finish it.'

"This is the man who killed your leader Arno," Moctu shouted, pointing to Zeke. At the mention of their former leader's name, the chants died to almost nothing.

Zeke's nose was seeping blood into his beard, but he smiled as if he hadn't been hit. Zeke motioned to the surrounding men encouraging their chant, then without warning, turned and rushed at Moctu again. The rush caught Moctu off guard, and the two men locked arms on each other's shoulders, trying to throw one another to the ground. Face to face, Zeke snarled something, and he began to move his right hand downward to get to the damaged ribs. Moctu's Krog training kicked in, and he head-butted Zeke's left eyebrow ridge hard, stunning him. Moctu slipped his right leg behind Zeke's right and pushed, toppling him.

But falling backward, Zeke managed to grab the back of Moctu's neck, and he pulled him down with him. Moctu landed on Zeke's chest with his sore left side, knocking the wind from the big man, but sending lightning bolts of pain through Moctu. Both men scrambled to their feet, each throwing tired punches that did little damage.

Once again Zeke rushed him, and Moctu delivered a powerful punch to the warrior's mouth, knocking part of a tooth out, but slicing his middle finger to the bone. But Zeke managed to get Moctu's head locked in the crook of his elbow next to his side. He squeezed hard, hoping to choke Moctu to death.

Can't breathe—gonna break my neck, Moctu worried.

But again, Moctu's Krog fight training surfaced, having been drilled in by months of practice during the long winter so far back in time. He moved his hip forward against Zeke's hip. Moctu circled his right arm around Zeke, nestling the side of his wrist just below Zeke's ribs. With his left hand, he pulled hard on his right, and he heard Zeke cry out. Moctu smiled, knowing from experience how much it hurt, feeling like you were being cut in half.

Zeke's headlock loosened as he used his right hand to flail away at Moctu's head which was still restrained by Zeke's other arm. It hurt, but Moctu just pulled harder, and Zeke's grip on his head gave way so that he could use both hands to try to dislodge Moctu.

A powerful blow from Zeke's right hand stunned Moctu, and he finally released his clamp on Zeke. As Moctu pulled his head up, Zeke continued, throwing punches with both hands, which Moctu only partially blocked. A loud cheer erupted from Etseh, and puzzled, Moctu glanced up to see his people pointing westward. On the distant horizon were two columns of dark smoke.

Chapter Forty-One

"Wait," Moctu yelled, holding both hands up.

Both fighters were battered and exhausted, and thinking Moctu might be giving up, Zeke paused.

Moctu kept one hand up toward Zeke while he looked to Creed.

"The fight is to the death," Creed said, "But perhaps—"

"No more have to die," Moctu said. Even to him it sounded weak—like he was giving up—so he quickly added, "I don't want to have to kill this man." He motioned for Creed to translate to Zeke and the others.

On hearing Creed, Zeke smiled, spit blood from his mouth, then looked to his men and laughed. They laughed with him.

"Ask Zeke what he loves more than anything," Moctu said to Creed. Then louder he said, "Ask him."

Creed frowned and his eyes narrowed. "What he loves?"

"Ask him what he loves more than anything," Moctu repeated. "Just *ask* him."

"Uh, the Nerean wants to know what…what you love most… more than anything," Creed said in their language to Zeke while wrinkling his nose in confusion.

Zeke's head flinched back slightly, and he looked to his men who were equally puzzled. Then he laughed and said in a loud voice in Nerean, "Women—your women." His men laughed with him.

"Besides women," Moctu said, louder still. "What do you love the most?"

Creed, now fascinated, repeated the question to Zeke.

Hearing that question, Zeke was sure he'd won the fight. Moctu was bargaining, and he'd give anything to stop the fight and not give up Nuri.

"Maybe you keep Nuri, but give us four women," Zeke offered.

Moctu sighed in frustration and rubbed his forehead.

Through clenched teeth he said to Creed, "What does he love most about his life—his way of life?"

Creed translated, and Zeke frowned and pursed his lips. Then he shrugged and said, "The sea…our boats."

Moctu let out a relieved breath and nodded. "And we have your boats now. You…you have no boats."

Creed's jaw dropped open, and he said, "You have…" Tilting his head at Moctu, he said slowly, "You have our boats?"

"Yeah, and we'll burn them all if we need to—if this attack continues…if you don't do what we want." Moctu pointed to the western sky. "See those two columns of dark smoke? That means we have your boats."

Creed's eyes widened, and in a hushed tone he said, "Don't burn the boats." He shook his head, confused at what to do next.

"Creed," Zeke yelled, putting his hands palms up, wanting an explanation of what was being said.

Creed swallowed hard before translating to Zeke. Halfway through the translation, Zeke's head bucked back, and three of his men arose from their seated positions to check the western horizon.

As Creed finished the translation, Zeke's lips pulled back, baring his teeth, and with a guttural roar, he charged Moctu, his head down. Moctu was ready for it, and he brought his right knee savagely up into Zeke's face hearing and feeling the nose break. Even so, Zeke's rush bowled him backward, and he landed hard on his back with Zeke on top. The pain was unlike anything Moctu had ever experienced, nearly paralyzing him and knocking his wind out. Gasping for air, but getting none, and with wicked pain exploding from his ribs, he rolled in the mud with Zeke, both trying to gain the advantage.

Zeke ended up on top with Moctu's knees drawn up, protecting his midsection. Both battered each other with fists to the face and neck. Remembering another Krog move, Moctu slipped his legs around Zeke's waist, locked his ankles and squeezed with all his might. Zeke landed a powerful blow to Moctu's chin that rocked him, but continually constricting his legs tighter, Moctu kept punching back at Zeke's face, aiming for his bleeding, destroyed nose.

When Krog men used the leg squeeze, their tremendous leg strength could incapacitate an opponent and end a fight. Moctu's legs were almost as strong, and Zeke groaned and gave up hitting him, using his arms instead to try to disengage. Moctu remained clamped on Zeke's waist even while the larger man struggled to his feet. Zeke began bucking, slamming Moctu's back on the ground, each time sending tidal waves of pain through his core.

Moctu gave up his leg hold, and Zeke withdrew a step, allowing Moctu to scramble to his feet. Zeke's face was a ruin of blood and gore, hardly recognizable as human. Moctu wasn't much better, bleeding from his eye, nose and ear. His ribs had hurt so much and for so long that he was almost numb to the horrendous pain.

Can't breathe…can't get air, he thought backing up two steps.

"You…" he pointed to Zeke, but couldn't get enough air to continue.

From under his waist clothes, Zeke pulled a knife, and the two men circled one another.

"Non—no labana," Creed and several others yelled.

"You…" Moctu found a little air. "You killed Arno…from behind…" Moctu gasped trying to get enough air to finish his sentence. "And now, you cheat…in this fight."

Creed translated, and the outcry of the surrounding men increased in volume.

"You've lost," Moctu continued as he warily watched Zeke and kept his distance. "We have…have your boats. Killing Arno…that didn't work. Your men…they can see you're a cheat."

As Creed translated the last word, Zeke rushed at Moctu and slashed a wide arc with the knife. Although Moctu swiveled away, his left arm was sliced deeply just below the elbow.

Nuri and the other Nereans watched with horror as Moctu spun away from Zeke several times narrowly avoiding the blade.

"Stop!" she and Elka yelled at the same time. Alta, Zaila, and all the other Nereans were screaming.

"Stop. Stop!"

"You cheat!"

But as Moctu dodged one more time, Zeke caught him by the hair and pulled him to the ground.

As he raised the knife, there was a 'thud', then another, and Zeke remained motionless. Then his knife hand came down as he fell forward on Moctu, two arrows protruding from his back.

Chapter Forty-Two

"We believe you now," Creed said. "Many of us um…well, didn't trust Zeke much, and what he did today…" Creed spit to the side. "That dishonors us all."

Zeke's lifeless body had been pulled from atop Moctu and dropped to the side. Two Sea People arrows, one from Creed and one from the huge, beardless warrior, had been extracted from his back. Now, most of the Sea People warriors stood, many with their heads bowed.

Creed swept his hand along the line of men. "These men lower their heads—not to Zeke—but to you as honor. They uh, we feel…" He paused, searching for the right word. "Shame."

Moctu's rebroken ribs were no longer numb, and they sent electric pulses through his chest if he moved wrong. Still gasping for breath, he managed to wheeze out, "Who'll be your new leader?"

Creed allowed himself a thin smile, then shrugged and pointed a thumb toward himself.

"What about…about Zeke's sons and friends?"

"Dead. Two sons dead. Nerean arrows." Creed looked solemnly into Moctu's eyes. "No more war." He shook his head and looked down. "No more war." After a long moment, he once again eyed Moctu and said, "Peace. Peace forever. No boats burn?"

Moctu studied Creed and considered the question.

The Lion warriors are coming, and we don't even have enough food for a moon, he thought. But the Sea People seem to be friendly now. And I trust Creed. Not like Zeke—never trusted that guy.

"How many warriors did you lose?" he asked finally.

"Four dead, including Zeke. Seven wounded but they probably recover," Creed said.

Moctu nodded then said, "Your warriors…Zeke, he caused my people…harm…great harm. You give us the bows of your four fallen men…" He paused, thinking and catching his breath. "You give us their two boats and three skins of dried fish. And you and your men swear to the Spirits no more fighting."

In a deeper voice, he said, "As witnessed by the mountains and rivers, the sky, the wind and the clouds," Moctu stared into Creed's eyes. "…and the greatest sea, there will be peace between us. If the peace is broken by anyone, those people will be cursed by all the Spirits, and they and their families and friends will die. Peace and friendship only. Say that to your men, and if they agree…" He pursed his lips then nodded. "Then that's good. That's all. You all agree, and no boats burn."

Creed smiled, relieved at the terms, knowing they were giving up nothing they needed other than a little fish that could be replaced. He nodded solemnly, then translated Moctu's words to his men. All put their fists to their hearts and nodded. Creed turned back to Moctu, and he put his fist to his heart. "They have all sworn peace and friendship, and they realize the penalty for breaking the peace will be a curse by all the Spirits on them and their loved ones. I, too, swear this. No more fight between us."

Moctu smiled and took a deep breath, wincing at the pain. He slowly and carefully extended his hand to Creed who clasped it warmly.

Creed moved to embrace Moctu, but quickly thought better of it and stopped. Instead, he again put his fist to his heart and dipped his head. "You are brave warrior, good fighter, great leader," he said.

Moctu nodded, then asked, "What will you do now?"

Creed nodded. He looked skyward then said, "The night has begun showing us the stars of the Seven Sisters. They've arrived, and now is the best time for us to move north. Less storms, and the boats can make it across big waters safely. We explore to the north when we get our boats back."

"I'll send word for your boats to be returned," Moctu said. "Except for the two that we're keeping. You'll have your boats soon."

Creed blew out a breath and smiled. His head sank in relief and appreciation.

"Thank you."

"I have one request for you. If you come back this way, I hope you, personally, will visit and tell us what you've seen and discovered. If you decide to stay up north, please send us word. Send someone to tell us what it's like to the north."

"That's a wise request," Creed said. "If we find what we expect… what we want—good lands with good fishing and hunting—we'll be sending a few men back to tell our people to the south." He smiled broadly. "We'll want them to send more people—especially women. I'll tell those men to stop here and describe all we've seen."

Moctu smiled and shook Creed's hand. Then he turned and slowly made his way up toward Etseh. Looking to the barricade he saw his family and friends expectantly peering down at him.

When he smiled and held a fist up, wild cheers broke out.

Moctu's smile broadened, but as he trudged on, it faded. "Now, for the Lion People…and Devu," he muttered.

PART TWO

The Lion People

Chapter Forty-Three

THE OLD STORYTELLER PAUSED her story and put a small log on the embers of the hearth fire. She blew out a long, weary-sounding breath, then drank from a skin of water. With a shaky hand, she checked an engraved bone about midway along the length of her necklace.

"I think that's all for today," she said to the adolescents gathered around her.

"No," two of the boys shouted in unison.

"What happened with the Lion People? At least tell us about that," the girl said.

"No, please don't stop."

"We're being good," the youngest boy whined. "Don't stop… please."

The silver-haired woman looked for a long moment at the youth, a stout boy of ten or eleven summers. Enthralled, he was leaning in toward her, and his eyes glowed with eagerness.

The furrowed lines on the woman's forehead softened, and she smiled and said, "Okay…a little more."

The young ones all whooped with delight and quickly settled back in their places.

Checking her bone fragment again, she murmured, "Oh yeah, this part's…"

The old woman didn't finish the sentence, and for a moment she stared off to the side, lost in thought.

The four youths watched mutely, wide-eyed with anticipation.

The woman finally nodded her head and looked back at her audience with a faint smile.

"The next part goes like this…"

Chapter Forty-Four

"LESS THAN A DAY away." Itzal reported to Moctu. "Almost three hands of them, and some are carrying clubs."

Nuri and Elka were in the hearth listening, and Lehoy was there, too, very interested and worried as a war band from his former tribe approached.

Moctu nodded. "So, they're warriors for sure—don't need clubs for hunting. How about bows? Are they carrying bows?"

"Looks like it, but none of them were strung, so they're still far enough away that I couldn't tell for sure. What they're carrying on their backs could be atlatl darts or bows—maybe both."

The Lion warriors were coming from the east, but Moctu looked to the west hoping to see Palo and Nabu returning. Two days before, the Nereans had sent up columns of dark gray smoke from three hills around Etseh. That was the signal that the Sea People warriors were leaving, and Palo and Nabu didn't have to burn the boats.

"They should get here in a day—Palo and Nabu." Thinking out loud, Moctu continued. "That'll help. And they're probably bringing four more good bows with them that they'll get from Creed. But with me, Zori, Indar and Alfer unable to pull bows, the Lion warriors still outnumber us."

"But we've got Lehoy, who's worth any three of them," Moctu went on, patting Lehoy on the shoulder. "You going to have any… misgivings about fighting members of your former tribe?"

He smiled at the well-built man next to him, who sported two long braids of dark hair which now had some strands of gray running through it. Lehoy still wore a necklace with a huge lion canine which marked him as one of the Lion People's greatest hunters. Years ago, he'd met and mated the green-eyed beauty, Avi, and he'd lived with the Nereans ever since. Lehoy had become one of Moctu's best friends and most trusted advisors.

"I've fought with warriors of my old tribe several times as you're aware," Lehoy said in a quiet voice. "I don't like doing it, but I'm a Nerean now. I have two Nerean sons," he said referring to his six and seven-year-olds. "And another child on the way."

Moctu smiled, then looked to his mate. "And you've gotten good with a bow, Nuri. Pick out one you like from the new ones that Palo brings. Elka, you already have a bow you're happy with, right?"

When Elka nodded, Moctu smiled at her and said, "You hit one of the Sea People coming up the hill in the leg. Even with him zigging and zagging. Good shot."

Elka smiled at the praise and cleared her throat. "Um, Itzal, was there a tall man in the group of Lion warriors? One with dark hair pulled back, broad shoulders?"

Nuri and Moctu looked at Elka with eyes narrowed.

Itzal considered the question. "Still pretty far away, but probably half of them could fit that description. Why?"

"Oh, nothing. Just...curious." Elka looked down, avoiding eye contact with everyone—especially Nuri. Her cheeks felt hot, and Elka rubbed her eye, pretending there was something in it. Why does Pele make me feel this way, she wondered. I don't even know what I feel. Part of me wants to kill him. Probably should have when he was hurt and I had the chance. But it felt wrong. Felt wrong back then. He'd just saved my life and Misha's.

But thinking of his tunic—the one she'd found in his pack—the one he was wearing when he'd earlier attacked her—and all the other evidence that he was one of the intruders, Elka's teeth clenched, and she shook her head. If he's in this group of warriors, I want to be the one who kills him. I *will* be the one who kills him.

Nuri had studied Elka ever since the girl lowered her head, and she knew Elka had an important, very private struggle going on inside.

“I think we should try to talk with them,” Moctu said. “Probably nothing’ll come of it, but we’ll get to see what weapons they have and maybe learn some things.”

Elka looked up. “I should go too,” she blurted.

When everyone looked her way, she stammered, “I…just…I can recognize…see if the ones who attacked us—if they’re part of the group.” Elka felt her face flush, and she scowled, trying to hide it.

But Moctu nodded and said, “It’ll be good for them to see that even our young women carry bows—maybe make them think twice about coming farther. And I’ll be carrying a bow, even though I can’t pull one right now. They won’t know that.” Facing Lehoy, he said, “I think it’d be good for you to come, too, if you’re willing.”

“Wouldn’t miss it,” Lehoy said.

Chapter Forty-Five

"There they are." Moctu said as he, Elka and Lehoy saw the Lion warriors emerge from a tall thicket of scrub brush.

The three were on a low knoll behind two large boulders and some small beech saplings that surrounded a weather-beaten, low oak. Moctu was still breathing hard having struggled to keep up with Elka and Lehoy. The pain in his sides didn't allow him to take full breaths.

The Lion warriors were just two bow shots away. "Two hands plus four—that's…that's all of the ones Itzal saw," Moctu said, sucking in another breath. "And they're carrying bows. *Crap.*" He frowned and paused to consider.

"I'm going to wave this branch in the air. Get ready to fire an arrow, then run if they attack us."

Elka busily studied the large group of warriors and was relieved not to see Pele among them. Why am I relieved? she wondered. "But they have bows," she murmured to herself. "Pele must have taught them how to make bows."

After waving the branch of truce and getting a similar wave in response, Moctu's group moved a few paces down the hill toward the Lion warrior party. A group of three warriors moved halfway up the knoll to meet them.

"That closest one is Gelari," Lehoy said. "He seems to be in charge. I see Naramsin's scarred face. He's another, older warrior, down in the main group. I don't recognize these other two young ones coming with Gelari."

Moctu watched Elka fit an arrow onto her bowstring but shook his head indicating she shouldn't go farther.

"That's far enough," he called. "Being in Nerean territory wearing war paint…that's very dangerous," he said loudly. "Good way to get killed."

Gelari smiled and said, "We're just following two deer we wounded."

"I'll pretend that you believe what you just said," Moctu yelled, smiling back. "If we find any deer with your arrows, well, we'll leave them at the border…the stream where it broadens." After catching his breath, he continued, "Any further moves toward us, and my people will be shooting at you from all sides."

Gelari's two companions quickly looked to both sides, but Gelari stared ahead. He scratched his short beard, then pointed at Elka and, giving up pretenses, he said, "Alright. Just give us her, and all this goes away. We don't have to be at war. Devu says—"

"Nothing but turds come from Devu's mouth," Moctu interrupted. "Think how many of your people have died because of him."

"You don't have to be another one," Lehoy said loudly.

"We're not going to leave this territory without her," Gelari said.

"Then you won't be leaving this territory, and your bodies will rot where they fall," Moctu called, his face reddening. He motioned for his group to back away, and Gelari did likewise.

Some distance away, on a farther hill, they could see that the Lion warriors were making camp.

"Glad they're not chasing us," Moctu wheezed. "I was dreading running for it. I can barely keep up with you two."

"Elka and I knew that if we had to run, we just had to be faster than you," Lehoy joked. But in a serious voice he said, "But they all had bows. That's not good. How'd that happen?"

Elka knew the answer, but she kept it to herself. Through gritted teeth, in a low voice she snarled, "I'll kill him. I'll kill him."

Chapter Forty-Six

"AT THIS RATE, WE'LL be eating nothing but soaked acorns within half a moon," Moctu said after Ono showed him the remaining dried meat provisions.

"Yes, with the cold weather coming on, this has all the signs of a starving winter," the elder advisor replied, his brow wrinkling even more than usual. Besides Jelli, the medicine woman, Ono was the oldest member of the tribe.

Moctu studied the old man for a moment—a man he'd known his whole life, and one who'd been an advisor to him or other leaders as long as Moctu could remember. He'd not always agreed with Ono, but his counsel was typically wise and valuable. The old man was tall and so reed-thin that most of his bones bulged through his dark skin. His untended hair was completely gray and scraggly and hung from the sides of his head. Ono's bald top was marred by scars, moles and scaley patches.

Ono won't survive a starving winter. He'd be the first to go, Moctu thought as the old man turned and limped toward the other storage niche which held even less dried meat.

"A lot of us will starve if we don't hunt soon," Moctu agreed. "But the Lion warriors stay just close enough to worry us and keep our hunters from heading out."

The two arrived at Moctu's hearth and Nuri ushered Ono in, encouraging him to sit on some furs that covered a boulder near the fire.

"Yeah, it's clear that's their plan. Because they have more people—more warriors and hunters than we do, they can threaten us with half their warriors while their hunters gather meat for the winter. All the while we're pegged down here, protecting Etseh."

Ono nodded. "I'm sure the Lion People shelter is lightly guarded, but we can't risk sending warriors there while so many Lion warriors are near Etseh."

Nuri sat with the men, and the three remained silent watching glowing sticks in the fire give way and drop into the pile of embers.

It was Nuri who spoke first. "We have to attack them." She paused, expecting disagreement. "We do—we have to. Otherwise, in two or three moons, it'll be death for many of us. And we know the area around here better, and they're not expecting us to leave our protected shelter. We can hit their camps then run. Then hit them again."

Nuri carefully watched the faces of the two men, but both remained transfixed, staring into the fire. Just when she was sure they had dismissed her idea, Moctu began nodding. He pursed his lips, and looking toward Ono, his palms opened, and he said, "She's right. I don't see another way to stop this. We fight. We attack them, and either we or they die now, or it's just us…us dying later in the middle of winter."

Ono kept staring into the fire, but his head began to nod. "I agree," he said finally.

Energized, Moctu stood and hugged Nuri. "Can you get Nabu and Palo?" he asked, referring to the other elders, "Let's plan this thing. You, too, Nuri. We'll need you here too. I think some of your traps will play a role in this."

Later, when Nabu and Palo were settled on the rock next to Ono, the hearth became quiet as all eyes fell on Moctu.

"We're going to attack the Lion warriors," he started. "Can't just sit and wait here while our food stocks disappear. I think we need to put the yew poison on all our arrow tips. We'll have two small teams that hit them separately—especially in the evenings when they're making camp. Hit them, then run. We know the terrain better and can disappear into the forest or brush. We can find our way in the dark, and they can't. And Nuri can have a group that makes traps along the routes they'll likely take when they chase us.

He paused, then said, "What do you think?"

'How big will these two teams be," asked Ono. "We can't leave Etseh unguarded."

"Three or four in each—that seems best to me," said Moctu. "And they don't have to be out there at the same time. We can keep plenty of bowmen…" He looked at Nuri and added, "and women in Etseh for protection."

Nuri cleared her throat and said, "I'll need to be out there to guide our warriors away after the attacks so they don't step in the traps," she said.

After a long moment of silence she added, "And I'll be out there anyway, building the traps."

Moctu frowned, not wanting to expose his mate to danger, but then he nodded and said, "Can't argue with that. I think—"

"I volunteer," Nabu interjected. "And Palo will want to go too—I'm sure."

"You two are the fastest in the tribe, so that'd be great," Moctu said. He lifted his elbows away from his sides and winced. "Wish I could go too." He shrugged. "But my ribs are healing—it won't be long."

"Moctu, I'm going to make another suggestion—one that'll trouble you, but hear me out," Nuri said in a quiet voice. "Elka should be on one of the teams. She sees the best in the dark, way better than anyone else, and her wolf, Misha could be a big help too." When Nuri saw that Moctu was unconvinced and beginning to shake his head no, she hurriedly continued. "There are people in the tribe that wrongly blame her for the Lion warriors being here. But Elka sometimes…well, she sometimes even blames herself. At least she questions it. Having her out there helping the teams would solve all that. At least it would help."

Moctu frowned, but Ono and Nabu began to nod, so Nuri kept at it. "And I know Elka would love the opportunity. She'd be great—a big help, and she'd love it."

Seeing the two elders nodding in agreement, Moctu finally pursed his lips and gave a curt nod.

Looking down at the ground he grumbled, "Okay."

Chapter Forty-Seven

Nuri shivered as she adjusted the skin that covered the hearth entrance. A frigid breeze was finding its way in through the spaces between the overhang's stone ceiling and the hearth covering. She paused to take in the fierce beauty of the storm. Towering, dark clouds of every shade of gray filled the northern and western skies, and they were headed this way. Although it was morning, the darkness made it feel like evening had arrived. Lightning flashed every few heartbeats, some in the sky where the clouds were fighting among themselves, but most of it hitting the ground with powerful crashes that seemed to shake the earth. Nearly black sheets streaked down from the clouds to the northeast, and Nuri could tell it was raining hard there.

"Serves them right," she murmured, smiling at the certainty that the Lion warriors would suffer through a miserable day.

Moctu was out, either with Palo or Nabu, and Nuri had just finished wedging small wads of skins in the largest spaces to block off the cold drafts when Elka and Misha returned from lookout duty.

"It's awful out there," Elka said, "And it's going to get worse."

"I know. It's the worst storm of the season so far. Looks like we'll get sleet. I'm freezing—been trying to block out the drafts."

Elka sat down by the fire, but she got up almost immediately and rushed outside, where she threw up just past the dripline of the overhang. Misha devotedly followed her, and as Elka returned, the wolf lapped at the small puddle of vomit.

"Come on Misha," Elka called irritably as she retook her seat by the fire. She saw Nuri eyeing her and said, "Second time this morning. Don't know why my stomach's so upset."

"You threw up yesterday morning too," Nuri said, feeling Elka's forehead. "No fever—that's good." After a pause, she said, "Poor girl—you've been pushing yourself too hard. Sit and rest." She added a small log to the fire and said, "Any other symptoms? Anything else going on?"

"Just tired," Elka said. "Really tired."

"Um..." Nuri said. "Elka, when was your last menses?"

Even before Elka began counting back, her eyes widened, and she looked up at Nuri, knowing the implication. "Well, let's see. It's been more than a moon, that's for sure. I...Oh, Spirits—do you think?"

Nuri's eyebrows raised, and she smiled and gave a little nod. "I think so." She sat by Elka and put her arm around the girl, pulling her close. "A lot happened out there while you were gone that you haven't talked about. I can tell that you have a burden...something's troubling you. Something big...something happened out there, and you don't have to carry it inside all to yourself. We love you. We love you so much, and we're on your side. I think it's time you told me what went on out there."

Elka's head sank, and she gave a little nod. "Can't believe I'm going to have a baby." She paused, putting a hand to her queasy stomach and shaking her head in disbelief. "I'd pretty much given up hope of ever having a mate, much less a baby. And this baby..." Elka stopped and looked directly into Nuri's eyes. "You really think I'm pregnant?"

Nuri's eyebrows once again rose, and she gave a knowing nod and a reassuring smile. "Pretty sure."

Elka smiled, then stared off into space. Nuri watched the smile fade into a worried frown.

"I thought I loved him, and I thought he loved me, but..."

As Elka stared at the wall of the hearth and shook her head, Nuri squeezed her shoulder and said, "But what dear?"

"Can men do that? Can they be so loving and tender and not really mean it? How can they...?"

Nuri shrugged, not wanting to answer truthfully and hurt Elka. "I'm afraid…" She began nodding regretfully. "Yeah, I'm afraid that some can."

The two women sat silently for a long moment before Nuri said, "But what makes you think he's…that he's like that?"

A tear streaked down Elka's cheek, and she looked into Nuri's eyes. "Because he's a monster. He's…" But Elka couldn't bring herself to say, "He's the man that killed Meko—your son, my brother." Instead, she said, "Let me first tell you that this man saved my life and Misha's."

Nuri hugged Elka, pleased that more of the story was coming out. "Well, he doesn't sound all that bad. How's he a monster?"

"He just saved me because I'm worth more to Devu alive than dead. Devu wants me for the sacrifice."

"So, this man, from the Lion People…" Nuri started.

"No, he's from the Sea People."

"Okay, but wait… So, the night attack two moons back… Who was that? Was it the Sea People, not the Lion People?"

"Uh huh…well, probably. I think Pele—that's his name—I think he's been with the Lion People before."

"So, this Pele—he was one of the attackers? What happened to the other guy?"

"I killed him. At least I hit him in the chest with an atlatl dart. Never saw him again. Just about has to be—"

"But you saw Pele with him?" Nuri interrupted.

"Didn't see him clearly, but it was him. His face was painted—red, black and white. Tall, broad shoulders. And his red tunic—with red and white beads. Wounded me with a poisoned spear," Elka said almost matter-of-factly. Then her blue eyes turned cold and flinty, and her lip curled as she said, "Bashed in Misha's eye. Almost killed her."

"Oh my—you poor dear…" She glanced at Misha and corrected herself. "Poor dears. We've all wondered how Misha got that…" Nuri squeezed Elka tighter. "Then what?"

"I was sick…and Misha was almost dead. Freezing…we…" Elka shook her head remembering that awful time. "Wolves—a big pack of them…Misha tried to fight but…" Elka brightened as she went on. "That's when Loebo came back," she said with a smile.

"Loebo? Your Loebo?"

"Yeah, Loebo remembered us—he saved us. Looks like he's the leader of the wolf pack. They left us alone after that." Elka's smile faded, and she shuddered. "But cold—so cold. We were freezing—both of us hurt so bad. Couldn't hunt. Could barely move. We were dying. That's when Pele came along."

Elka shrugged. "He didn't kill us. He could have." Elka pursed her lips, thinking. "Guess he wanted us alive—worth more to Devu."

"You didn't recognize him?"

"No, he wasn't wearing paint, and he'd hidden the red and white tunic in his pack. Gave us food, furs and medicine. Saved us. He was really nice." Elka's face clouded. "I should have figured… What's the likelihood of two different men like that out in the middle of nowhere?"

"But he saved you and Misha so…"

"Yeah, he was so nice. Took care of me. Taught me to use a bow. Seemed to like me." Elka stared into the fire, which was now just a pile of glowing embers. Her voiced cracked as she went on. "I thought a man finally…finally loved me." A tear beaded in her eye and rolled down her cheek which she quickly wiped away with the back of her hand.

Nuri squeezed Elka hard against her chest. "Don't you worry sweetheart. There's going to be someone for you. A good man who loves you."

The two were quiet for a long time as Elka composed herself.

"But now, he's teaching the Lion People to make bows, and they're coming to attack us," she said finally. In a detached voice she added, "I'll kill him. Should have done it before. But I'll kill him."

Chapter Forty-Eight

"Our first hit and run raid wounded a Lion warrior and caused a lot of panic," Palo said, reporting back to Moctu after three days out. "The second raid was…" He grimaced. "Well, it was a disaster. Their warriors were ready, and they quickly spread out around us, so we had to split up and run for it. I haven't seen Nabu or Elka and her wolf since."

"Oh crap," Moctu said, rubbing his forehead. "I was worried… this is exactly what I didn't want to happen."

"I'll bet they'll be okay," Palo said weakly. "They're both smart and crafty."

"Well, but Elka's pregnant too."

"Elka's pregnant?" Palo asked.

"Yeah, she made Nuri promise not to tell me before she went on the raid. Wanted to be treated normally. But that…" He paused shaking his head. "That was a big mistake. Crap, *crap*." He thought for a moment, then said, "This is going to kill Nuri. She had big misgivings, but she went along with what Elka wanted because the girl's so determined and stubborn. I'm just glad Nuri didn't go along with her—too much mud out there to make her pit traps."

Moctu paced while Palo bit his lip and stayed quiet.

"If they have her," Moctu said finally, "they'll leave for their shelter. Get her back to Devu for a sacrifice. If they have Nabu, or if they don't have either of them, they'll probably stay. We have to be ready to go after them if they leave."

"We'll need a good-sized group." Palo said. I'll round up Lehoy, Seetu and… Alfer—I think he can pull a bow now. And I'll go, of course."

"I'll be going too." Moctu said. "I can pull a bow far enough to…well, maybe not to kill, but I can hurt them. And, in a day or so, I'll be back to normal, and that's when we'll catch up to them if they've left."

When Palo narrowed his eyes, Moctu added. "I won't slow you down. Listen, if they've got Elka, I'm going, and Nuri's going to want to come too. So, check with Koby who's been watching from the high eastern lookout—he can see them when they get to the lowlands, if they're headed back east."

"And how about Nabu?" Palo asked.

Moctu frowned and blew out a breath. "I'm afraid if they've captured Nabu, he's not going to last long. They'll torture him..." Moctu pursed his lips. "Then they'll kill him."

"Well, we need a plan for that too," said Palo, his face pale. "Something quick. Can't just let him be slaughtered."

Moctu saw the anguish in his friend's face and knew Palo was feeling enormous guilt. He patted Palo's back and said, "If there's one person in this tribe that I would bet on to get away from those guys, it'd be Nabu. He's really fast and smart. But you're right. We need a plan to get him back if they catch him. Can't lose Nabu."

"I just keep playing it over in my mind—what I should have... what *we* should have done differently," Palo said looking at the ground. "The plan was to hit them and run, but...but I should have stayed and fought."

"No. Then you, or all three of you'd be dead. There's too many of them. Like you said, the plan was a hit and *run*."

Moctu and Palo were still talking when Koby stuck his head between the hearth entrance skins and said excitedly, "Nabu's back." His countenance clouded, and he added, "No word on Elka yet."

Chapter Forty-Nine

Elka carefully peered over the shallow ditch from between two young pines, then she touched Misha's head directing the wolf to complete quiet and stillness.

"He's coming this way," she whispered to herself and the wolf, and a wave of nausea swept her. The situation was eerily similar to that of several moons before when she almost stumbled into the campsite of her brother's killers as she was tracking them.

They were hunting me then, too, she thought, placing a hand on her queasy stomach. *But now I have a baby on the way.*

That was my first meeting with Pele, she reflected ruefully, *and it sure didn't work out well.*

Elka's thoughts were immediately directed back to the moment as the Lion warrior looked directly at the trees in front of her. Ducking her head down, she hoped he hadn't seen the slight movement. To the left side of her about a bowshot away, another warrior thrashed through the brush.

They're focused on me, she thought. *Spirits, I hope Nabu and Palo got away. There's so many of them!* Elka once again nervously fingered the arrow nocked on the string ensuring it was positioned right. *Only three arrows left, and there are at least that many warriors right around here.*

The Lion warrior nearest her was a big man and older than most of the others she'd seen. He wore the traditional necklace with Lion's tooth strung between two white shells, and it was a big tooth indicating he was an expert hunter and warrior. He had black paint surrounding his eyes which made him look ghostly, and the beard around his mouth was stained red with ochre, making it look bloody. The effect was creepy and unnerving.

"Here's a wolf track," Elka heard him exclaim in the Lion dialect, and others nearby answered back that they were coming his direction.

"Time to run again," she whispered to Misha. "But first…" She rose quickly and fired at the approaching warrior. He was too close to miss, but he ducked, and instead of hitting him square in the chest, her shot took him in the shoulder. He called out in surprise and pain as the shot spun him around and put him on the ground. Immediately there were answering cries from all around her.

"Crap, they're everywhere," Elka muttered. "This way Misha." With that, Elka charged through the brush directly away from the downed warrior. She ran for a few heartbeats with Misha behind and to the right of her, and then she saw movement in that direction.

"Whizzzz!" and arrow flew just in front of her eyes, and she felt the breeze from its passing. "Too close Misha—faster!"

Another arrow flew wide, and Elka zigged away from the newest threat. She and Misha pounded down a muddy ravine full of scraggly hornbeam and young maples, and they ran along it for several strides before beginning their way up the slope on the other side. Elka felt the hair on her neck rise as their progress slowed on the less-vegetated, upward slope, and their backs were exposed. She braced for the impact of an arrow and gritted her teeth, climbing harder and faster.

"Thank you, Spirits," she wheezed as they safely crested the ravine and began to run through a maze of young pines and maples that competed for the best sun. Once again, Elka felt nauseous, and she slowed to avoid throwing up, which would slow them even more. Even Misha was panting, but she ran, her shoulder occasionally brushing Elka's leg seemingly encouraging a faster pace.

"I know, I know," Elka said, "Just give me a moment."

Ahead and more to the right, she saw that the brush thickened, and they would find more cover there.

"Might be able to disappear in there," she said and veered toward the thicket.

That's when she heard it.

"Whizzzz—thud!"

"Ark-aik-aik -aik!" Misha yelped piteously.

Elka looked back to see her wolf dragging her right leg, an arrow protruding from just below the hip.

"No!" Elka screamed.

Misha limped toward her, mostly dragging her hind quarter with the arrow jutting hideously outward. She whined as she hobbled to Elka, then seemed to smile at her. "I can still go on," her eyes told Elka.

Elka fell to her knees, holding the wolf's head in her arms, giving up on escape.

"I'm so sorry Misha. I'm so… Oh my girl, Oh my sweet girl."

Moments later she was surrounded by Lion warriors, their bows drawn.

Chapter Fifty

"THEY'RE HEADED BACK EAST toward the Lion shelter," Zori announced breathlessly. "Ran…I ran here as fast as I could. There's some wounded…" The young man paused to catch his breath. "Some wounded being helped along—I didn't see Elka for sure. But…"

Moctu frowned at the youth whose arm was still bandaged from an arrow wound. "But what?"

"Well, there was an animal with them…a wolf. Looked like it was wounded."

Moctu's head sank. After a moment he looked up and said, "Thank you, Zori, you've done well." He paced a few steps then said, "Go tell Palo to round up Lehoy, Seetu and Alfer. We'll be heading out after them immediately. Hurry."

Zori left in a rush and Moctu closed his eyes. His mind flashed to a promise that he'd made long ago to his first mate, Effie, as she lay dying. With her last words, she'd said, "Take care…Elka," and he'd resolutely agreed.

"Oh, Spirits, please protect Elka," Moctu said. "Please keep her…" He shook his head sadly then put his palms upward. "…and my grandchild safe. Please." With that, he hurried to Avi and Lehoy's hearth, where he found Nuri seated with her arm around Avi's shoulder, trying to talk as the two young boys wrestled on the ground, giggling and shrieking noisily.

Avi yelled crossly for the boys to go elsewhere, and Moctu noticed her light brown hair—a rarity in the tribe—was tousled and dirty. Normally one of the prettiest women around, with fiery green eyes—another rarity—Avi looked a shell of herself.

Before Moctu could talk, Nuri said, "Indar is better, but…" She shook her hands. "But Avi's pregnancy," Nuri paused looking back at her best friend. "Well, she's worried she might be losing the baby."

Moctu opened his mouth to speak, but Nuri went on. Holding up a piece of mammoth tusk carved to look like a woman with all parts of her body oversized, she said, "We've been praying to the Earth Mother for Her blessing."

"I'm…I'm glad to hear about Indar," Moctu said. "And I'll ask the Spirits to keep you and your baby safe, Avi."

"Thanks, Moctu," Avi said sniffing and wiping a wet eye.

"Aaa, Nuri," Moctu continued hesitantly. "Looks like Elka's been taken."

"Oh no, no," Nuri said, putting her hand to her mouth and standing.

"I'm going after her," Moctu continued. "And Avi, I've asked Lehoy to come too—along with Palo and Seetu."

"I want to come too," Nuri said quickly. "Oh Spirits, why did I let her go?" Turning to Avi she said, "We'll be praying to the Spirits for you Avi, and I'll be back as soon as—"

"You go now—I understand," Avi said, waving her off. "Thanks Nuri. I'll be alright."

As Moctu and Nuri walked away, he said, "I don't think it's a good idea for you to—"

"Don't try to talk me out of this, Moctu. I'll never forgive myself if anything happens to Elka. And her being pregnant too. Oh, Spirits."

Moctu put his hand to Nuri's back and said, "It's not your fault. I don't think you could have stopped her. She already went off once on her own. She's stubborn and strong-willed. And she's tougher than any young man her age." After a pause he added, "And she's got Misha."

"I know all that," Nuri answered hotly. Then she slowed and put her hand on Moctu's arm. "I'm sorry. Not mad at you. Mad at *myself*." She squeezed her forehead, then looked up at the horizon where the sun was well into its dive toward the western hills. She held her hand out and said, "About two hands of daylight left. I'll be ready to leave in one finger of time."

The group assembled quickly and included Palo, Lehoy, Alfer, and Nabu who subbed in for Seetu who wanted to stay with his injured mate, Leuna. As they always did, Moctu's eyes narrowed as Alfer approached. Long ago, the older man had repeatedly beaten his mother Alta while Moctu was away. It hadn't happened again since Moctu threatened him, and Alfer was Elka's step-grandfather, so maybe he'd be an asset to the group.

The group set a fast pace, jogging, then walking, then jogging again. Although Moctu struggled and his ribs burned during the jogging, he found he could recover during the walking stretches. By sunset, the band was within a half day of the lowlands.

"At this rate, we'll catch up to the Lion warriors in…about a day, maybe a little more," Lehoy wheezed after a mostly uphill jog. "Then what?"

"Then we see if they have Elka, and we make plans…" Moctu shrugged and worked to catch his breath. "Based on what we see."

Chapter Fifty-One

"I'll go faster—and willingly if you let me treat my wolf," Elka said as she labored to carry Misha, who weighed about what her adolescent brother had. With each step, the arrow swayed wildly back and forth, like a sapling in a windstorm that changed directions with every heartbeat. "Please—she's hurting! At least let me get the arrow out."

The nearest Lion warrior, a chubby youth, said nothing but kicked the girl, almost toppling her. An older warrior, who had an ugly, arcuate scar that ran across his left eye, leaving the eye itself gray and useless, barked at the young man.

"Enough of that Beotuk. The Spirits are with Sleeps With Wolves and allow her to speak to and control animals. And that wolf—he has the same eye as me. That means something—something powerful."

"Yeah, but she's the one who shot Gelari," the chubby warrior complained.

"Beotuk!" the one-eyed man bellowed, looking sternly at the youth.

"Yes, Naramsin," the boy responded meekly.

Naramsin pointed eastward, and to Elka he said, "We'll stop for a short while at that knoll with the small pine. What will you need to…to heal the wolf?"

Elka had learned healing from both her grandmother Rah and Nindai, and she could sing each of the long medicine songs of the Krog and the Nereans. The songs helped the healers memorize the plants, preparations and techniques involved in making the many potions and poultices that could cure the sick or injured. Knowing the overlapping but different medicine songs in the two different languages was something neither Rah nor Nindai had mastered.

"I'll need a fire, some water, and mandrog root or belladon," Elka answered Naramsin. "And two long poles for a travois. If you don't have mandrog or belladon, I can probably find some on the way." After a pause she gave the man a curt nod and said, "Thank you."

"We don't have any…either of those plants," Naramsin said.

"Mandrog's easy to find right now. This is when it blooms—bright purple flowers. That might be some over there," Elka said pointing.

"You stay here," Naramsin commanded. "We'll bring it to you."

"It's the root I need. The best mandrog looks a little like a person's body."

Naramsin dispatched two men to dig up the plant with the purple flowers, and sure enough, they brough back a root that started as a thick, bulbous tuber that had two tapering legs at the bottom.

"Perfect," Elka said. "That'll work well."

From a small side pouch Elka extracted the last piece of her caribou jerky, and on a flat stone she beat it into tiny shreds. Next, she took the tuber and tore off the bottom half, and she beat it into a wet pulp that mixed with the jerky. Careful not to get too much on her fingers, she scraped about half of the mixture onto a flat cobble, and she offered it to Misha.

There was enough caribou—Misha's favorite food—in the mash that the wolf quickly gulped it down.

"That should put her to sleep quickly. Then I can get the arrow out and tend her wound."

In short order, Misha's head began to droop, and her slack tongue lolled to the side. Elka pushed the wolf's tongue into place, and with a long thin piece of rawhide, she tied Misha's jaw closed.

"Don't want her to wake up while I pull the arrow out. It's really going to hurt," Elka muttered to herself as much as the interested group of men that was now watching her.

"I wouldn't pull an arrow from a live wolf, even with the mouth strapped shut," Naramsin remarked to murmurs of agreement from the group.

At Naramsin's orders, one of the men built a fire, and Elka located a long slender stone and placed it halfway into the embers. She prodded Misha to ensure the wolf was asleep, then Elka stood and positioned herself to extract the arrow. With a foot softly placed on the wolf near the arrow, Elka grabbed the shaft and prayed.

"Spirits, please don't leave the arrowhead inside." With that, she pulled hard on the shaft, and out came the arrow—including the flint tip. Misha barely stirred.

"Oh, thank you Spirits," Elka said as several onlookers murmured approval. "The tip is out and she's not bleeding too much."

Elka turned and with cold, hard eyes she watched the group as she very deliberately broke the arrow in half. After a moment glaring at the group, Elka turned to the fire, and using a thick wad of skins, she picked up the red-hot stone and held it to Misha's oozing wound. Even in her sleep the wolf yelped and tried to edge away, but the cauterization was over quickly, and it served to staunch the bleeding. Elka left the wound unbandaged, knowing Misha would chew any wrapping off, and she would want to lick the wound clean.

Elka saw Naramsin look at the sun, and she heard him direct a young warrior to run ahead to the Lion shelter to alert them that they were returning with Elka. At their current rate, it would be several days for the main group to reach the shelter, and Elka knew Naramsin was anxious to get moving again.

"I'll have the travois built quickly," she said. "And we can get going again." As she strapped the poles together, she muttered quietly, "Wouldn't want to be late for my sacrifice."

Elka paused to look at their back trail but she saw no movement in the distance. She bit her bottom lip then murmured, "Is anyone coming for me?"

Chapter Fifty-Two

The lowlands were marshy in places, and the ends of the travois sank deeply into the mud. Some of the warriors stacked on their packs and sleeping furs, increasing the load. With her food gone and none offered by her captors, Elka found it increasingly difficult to go on. Although a cold wind blew, the sun was out, and perspiration dripped down the sides of her face.

"Getting closer and closer to…" She shook her head and wiped sweat from her eyes. "To my…death."

The word had such finality that it took her breath away. She thought of Moctu who always said, "Live your best life, and die your best death."

"But have I lived my best life? That was it?" Elka murmured. She snorted bitterly and repeated it louder, "That was it? Yeah, I had love in my life—for a few days." She scowled at the thought of Pele. "And now I have a baby. But it's a baby that'll never be born." A tear welled in her eye as she realized that she would never see her baby, never know if it was a boy or a girl. Never know the child's personality.

She angrily wiped the moisture away before the tear fell. "I won't let them see me cry," she hissed through clenched teeth. "I've had a good life—good parents, grandmothers…" Elka looked back at Misha who was still strapped atop the travois. The wolf was licking her wound—and recovering nicely. "And who else has had the friendship of wolves?"

The last thought allowed her a thin smile. "I have had a special life at least. That's something, I guess."

Ahead, she saw Naramsin stop and survey their back trail. He seemed satisfied, and he moved toward her.

"We're making good time, all things considered," he said.

The two walked together in silence for several paces, almost amicably. It was surreal. This man who had been kind to her and Misha was now walking them to their death.

"What do you think will happen when we get there…to your shelter," Elka said finally.

"Well, the moon is slender, and Devu will probably want to wait until it's full, so…" Naramsin opened his mouth to say more, but then frowned and said nothing.

After a few more paces, Elka said, "Tell me of Pele. Is he still with you? Has he taken a mate?"

Naramsin glanced at her, and Elka felt her face heat up.

Immediately angry with herself for blushing, Elka scowled to mask it.

"He um…yeah, he has a mate now."

Elka stumbled, almost dropping the travois, but quickly regained her footing. Her head spun, and she looked away from Naramsin as she tried to compose herself.

Naramsin didn't seem to notice. "You know Pele?"

Elka paused, considering her answer. "Yeah, he tried to kill my family," she said finally.

"Ooh," Naramsin said. "So, you probably—"

"Yeah, I want to kill him," Elka said brusquely. As they walked along, she muttered to herself. "Should have. But now…" She shook her head.

"Devu chose Mia as Pele's mate," Naramsin said. "Pretty and a hard worker. She's a good mate for him. He's lucky—many men wanted her. Devu wanted to keep Pele—to have him stay at our shelter."

Elka's nostrils flared, and she fought the urge to scream.

Why do I feel like this? she wondered. Why do I care?

Elka's hand began to ache, and she realized she was gripping the travois poles too hard.

Desperate to alter her mood, she asked, "Was Pele the one who taught you to make bows?"

Naramsin nodded and said, "Yeah. We had gotten one from another trader. But Devu got him to teach us to make bows.

Pele…he's so valuable for our tribe. Very good with his bow, too. Better than anyone else. And his bow is way better than the ones we make. He's going to be one of the top men in the tribe."

"Well, good for you," Elka said sourly. She paused pulling the travois as a wrenching pain in her stomach staggered her. Elka bent to lay the poles down, but dropped them instead as the pain overwhelmed her. Startled, Misha shook herself free of the skins that bound her to the travois. Elka went to her knees, both hands on her stomach as nausea caused her to wretch. Her stomach was empty, so only yellow, acidic bile emerged.

"Oh, Spirits," Elka cried, wiping a strand of saliva from her mouth with the back of her hand.

As she knelt on the ground, exhausted and spent—both emotionally and physically—the Lion warriors gathered to watch. The wounded man, Gelari, was supported by two others, and he looked extremely pale and haggard. But he managed a weak smile and said, "Good, now *you* suffer some."

Through narrowed, bleary eyes, Elka peered up at him, but she had no energy to respond.

"Spirits," she whispered to herself, her lips barely moving. "Losing the baby. Please Spirits…please help." Elka shook her head sadly. "But no point really… Be dead in a moon." Her stomach turned, and she wretched again, but nothing came out. A deep, encompassing gloom overpowered Elka, but it was different from her earlier sadness times. Now, she had no thoughts of harming herself—she just wanted to be left alone with Misha and the baby inside her. Elka closed her eyes wanting only to sleep.

"Never felt like this…this awful," she muttered as a murmur swept through the warrior onlookers.

Elka opened her eyes in time to see Misha limp the last step to her and with a warm, raspy tongue, begin to lick her face.

Chapter Fifty-Three

"What's all this burned brush?" Palo asked looking north. "Never seen so much. Must have been a huge range fire. We need to stay in this unburned stuff to make sure we're not seen. And if we leave the cover to follow them, they're going to see us," Palo said.

"They're staying in the brush south of the burned area," Lehoy said squinting. "I only count two hands plus two—besides Elka. And one's wounded. Can't make out much else. Gelari might be there, but I can't see him. Elka's walking and carrying a travois. I don't think she's hurt, so that's good."

"Yeah, that's good," Moctu said. "Now we just have to get her away from a bunch of warriors who outnumber us two to one."

"They're less than a half day away," Nuri said. "It's cold, and they've been having fires at night. So, we can catch up to them after dark. Risky, but it's probably the best plan."

"And once we get there, what are we doing?" Alfer said scowling. "Besides getting ourselves killed."

"Well, they're less than two days from the Lion People shelter," Moctu said. "It's only going to get harder. Nuri's right—we've got to try the night raid."

"I want to get her back, too, but if we don't get some good hunts in, we'll all be starving in a few moons anyway," Alfer groused.

"Mmm, I don't see any other way," Lehoy agreed with Moctu, ignoring Alfer. "If they see us coming…there's just too many of them. And even if we got close, they'd likely kill Elka before letting us get to her."

"Okay, so we go after dark," Moctu said taking a seat beneath a particularly tall heather. Its pink flowers gave off a pleasant, sweet and musky scent, but after waving away several bees, Moctu decided to relax under a beech shrub instead.

The past two afternoons had been unseasonably warm, even though the nights were still cold, and they could see their breath in the mornings. While the other men in the group settled in shady areas, Nuri dumped her pack to the side and sat down next to Moctu. "How are we going to do this tonight?"

"I've been trying to come up with a plan that'll work," Moctu said, shrugging. "And it'll depend on how they make camp." He shook his head and blew out a breath. In a lower voice he said, "But it may be too risky. Much as I want to get Elka back, I'm not going to get us all killed trying. And maybe get her killed too."

"If we can take out the night guard—or guards," Nuri started. "And do it quietly, then I'm your best bet."

"I know," said Moctu reluctantly. "You're a Nightwalker—certainly the best in this group. You move through the night soundlessly. Better than anyone I've ever known. And it's what got you away from the Shiv back…way back then. But…Spirits, if I lost you…" He pursed his lips and shook his head. "First Meko… And now, if I lose you and Elka, I…" His head sank. "I don't think I could… Crap, I don't think I…" He didn't finish the sentence, and there was a long silence.

"But if we don't get Elka," Nuri said, "we *know* what'll happen. And I couldn't live with that if I didn't at least try."

"I know, we've got to try. I couldn't live with that either—not trying. So, I guess that's what we'll do. We take out the guards quietly, then you slip in, then we get away. Simple." Moctu snorted bitterly, then looked at the sky and said, "Spirits, please protect Nuri. Please protect us all."

The shadows gradually lengthened, and a chill set in. As the sun set behind the western hills, gold rays streamed between gray clouds edged with pink. The sight was beautiful, and Moctu's band took it as a good omen.

As dusk deepened into darkness, the small band set off in single file following Nuri. Directly behind her, Moctu could only make out motion, not form. He could hear faint rustling behind him from the others, but nothing from Nuri.

"She's so good at this," he marveled quietly, proud of his mate. "The only one who comes close to her as a Nightwalker is Elka. It's like they both can see in the dark."

About half the night sky was covered by clouds that blocked out stars and occasionally covered a crescent moon. Moctu frowned as he scanned the unveiled portion of the heavens.

The Seven Sisters have begun to get a little lower in the night sky, he worried. *Alfer's right, winter's almost here and it'll be a starving time if we don't have some good hunts. But first...this. Have to get Elka.*

Nuri took her time choosing the best paths for silent movement, and by the time the group was in position, the three fires of the Lion warrior camp had burned down to faint embers. It was time. Moctu's group set up in a semicircle on the western side of the camp, and all except Nuri had an arrow nocked on their bow strings. Holding just her keen-edged knife, Nuri moved in closer to the camp, her eyes scanning, probing for the location of the guard or guards. A cold gust blew past, causing her to shiver.

"There he is," she murmured to herself, spotting a husky, young warrior who looked to be asleep sitting against a sapling. "At least that's one of them." As she neared the youth, it became clear that he was asleep, and Nuri pursed her lips and paused to think.

Don't really want to kill him. He has a mother somewhere who'll be crushed. Could club him with a rock, but killing him is safer for me and our group. And he is *part of the group that's taking my daughter back for Devu to kill her...*

With that last thought, Nuri's lip curled, and her hand tightened on the knife. As she circled to approach the sleeping youth from behind, thoughts of her time with the Shiv came to her.

But Shaka-Nu was a full-grown man. Nuri slowed to reconsider, now almost to the boy. *Shaka-Nu was a huge Pale One, a cannibal. And this one...this one's just a beardless boy.*

"Beotuk," came a whispered call from a brushy area about a third of the way around the camp. "You awake?"

"Uh, oh yeah—I'm awake," said the youth getting to his feet.

Nuri's breath caught as she remained frozen at the periphery of the boy's view.

From slightly farther away this time, she heard the other guard say, "Okay then."

That's when the youth turned his head, and his eyes widened as he saw Nuri. She was on him as his lips parted to call out, but her cupped hand covered his mouth, and she pulled his back to her, drawing the knife hard across his throat. He only struggled for a couple of heartbeats before slowly going limp.

Nuri eased him to the ground, her right hand drenched in the boy's hot blood.

"You okay?" the other guard hissed.

Nuri's heart raced as she stared opened-mouth toward the voice. After several more heartbeats, in her deepest voice she whispered back, "Just peeing."

There was a chuckle, then sounds of the guard moving away.

Chapter Fifty-Four

Nuri waited for her hammering heart to slow and tried to breathe normally. Not far from her, there was movement in the camp, but it wasn't a man. It was a wolf, and the animal whimpered with pain as it repositioned itself next to a sleeping figure.

"Misha," Nuri murmured excitedly. "And Elka will be that person right there."

Backlighted by a dimly glowing fire, Misha didn't appear to be staked down, so the figure she lay against had to be Elka.

"She's only four of five paces in."

Nuri studied the sleeping figures, then looked toward where she'd last heard the second guard, then back to the sleepers.

"Only one sleeping warrior to get by. Do I try to kill the other guard, then go in, or just go in and get Elka now?"

Wiping her bloody hand on the front of her tunic, Nuri closed her eyes considering the choices.

"The remaining guard's alert—going to be tough to take him out quietly. Dawn's coming. So just go in now. Act normally—like one of the warriors who went out to relieve himself. Nuri nodded to herself, but a chill went down her spine causing her to shudder. She took a deep breath and murmured, "This is it. Spirits…please be with me."

Nuri could feel her heart thumping fiercely as she stepped out of the brush toward the sleeping figures. Walking casually, she moved past the one sleeping warrior between Elka and the edge of the camp. Halfway expecting an arrow to slam into her side, Nuri carefully made her way to the sleeping form that Misha lay against.

"It's Elka—thank you Spirits," Nuri murmured, relieved. As she began to kneel down beside the girl to wake her quietly, Misha growled loudly.

"Shh, shh, shh—it's me girl. Misha, it's me," Nuri whispered to the wolf, trying to get her to realize she wasn't a threat. But as she covered Elka's mouth gently with two fingers before awakening her, Misha's growl grew louder and more menacing still.

Elka opened her eyes, then blinked, her eyes widening in surprise. Then she nodded in understanding, and she tried to calm Misha with a quick pat to her head.

But it was too late.

"What're you doing?" the guard challenged from a thicket ten paces away. The sleeping warrior Nuri had bypassed started to get up, and she knew it was time to run. Nuri arose, pulling Elka up with her, and she sprinted toward the brush at the edge of the camp. An arrow whizzed just in front of her, and Nuri slowed, zigging to make herself a harder target. Close behind, Elka zagged the other way.

Nuri was almost to the edge of camp when a warrior tackled her and fell on her legs, pinning them to the ground. She slashed at him with her blade, carving a bloody line across his ear. But he was soon joined by another man who gripped Nuri's knife hand and held her neck to the ground choking her. Elka hesitated, then moved to help Nuri. But she was bowled over by a powerful warrior who crashed into her from the side. She went down hard on the ground, her hip landing on a large cobble. Limping behind Elka, Misha was there quickly and savaged the man's arm for several moments before she was thrown yelping to the side by another warrior.

Moctu's group could see nothing clearly, just commotion, and they held their fire, worried that they'd hit Nuri or Elka by mistake. Then a Lion warrior stumbled through the smallest of the three fires, sending glowing embers in all directions. Finally able to see—at least dimly, and charged with adrenaline, Moctu found he was able to pull an arrow back fully, despite his injured ribs. He let it fly, and it took the nearest warrior in the back near the left shoulder. The man spun and went down.

Some of the other men with Moctu likewise found clear shots and scored a few hits before there was a shout, "Shoot again and they die. Sleeps With Wolves dies. And this other one too. You hear?"

"Shit, they have them both?" Moctu muttered.

Elka struggled against the Lion warrior doing the talking, but he clutched her tighter, his arm around her throat. His tunic smelled moldy, and he stunk of sour sweat. Stronger than the young men her age, Elka pried his powerful arm up enough to yell, "Father, kill them all!"

"Let them go," Moctu yelled. "And we won't kill any more of you. We have you outnumbered," he lied.

The warrior yelled back, "That's a joke. We know you're fighting the Sea People, and you have to leave most of your fighters at Etseh. So, one more arrow from you, and I'll slit their throats." He paused, then said, "Devu won't mind too much if we bring Sleeps With Wolves to him dead."

Elka grimaced, stunned that the Lion warriors knew of the Nerean's fight with the Sea People.

How could they know that? she wondered. Must have been the Sea People's plan all along, and Pele told the Lion People, she realized. Her lip curled with the thought of that betrayal, and with a snarl, Elka went berserk, wildly struggling to free herself. She was giving the warrior holding her all he could handle until another Lion warrior punched her hard in the stomach.

It took her breath away, and Elka fought to get even a little air into her lungs. She ceased her struggling, and as she finally sucked in a tiny gasp, she worried, "What about my baby?"

"We've already beaten the Sea People, and now we've come for you. Let them go, and no more need to die. Do it now."

"That's not going to happen," the man yelled back.

There was a long pause as Moctu considered the standoff. His glance to the east showed a pale golden glow defining the low spots between the crests of the hills indicating that dawn was nearing.

"Won't be long before they can see better and tell how few of us there are," Moctu whispered to Palo. "Then they'll attack, and...I don't like our odds."

"Yeah, even with some wounded, they still outnumber us two to one. And they can always threaten to kill Elka or Nuri," Palo responded.

"Have to make a deal quickly," Moctu said, frowning as he once again eyed the ever-brightening eastern summits.

"Make a deal with you," came a loud call from the Lion warrior camp. "We'll give you this other one back. Gelari thinks it's your mate Moctu. She worth anything to you?"

"Give them both back, and we'll let you live," Moctu yelled, bluffing one more time.

"Here's the deal. We give you back your mate, and we keep Sleeps With Wolves. And she'll be treated well through the rest of this moon. No more attacks from you while we're traveling—on your honor." There was a pause, then the voice added, "And we won't come kill you all now."

Moctu forced a loud, contemptuous laugh, but he couldn't think of a way to respond. He nervously glanced at the eastern horizon where the sun's rays were streaming through the low spots of the hills and outlining even the peaks.

"Deal?" the man yelled.

Moctu closed his eyes tightly and grimaced before saying, "Deal."

Chapter Fifty-Five

"WHAT HAVE I DONE?" Moctu moaned to Palo. "That's my daughter...and my grandchild."

"Did what you have to," his friend responded.

"Hardest thing I've ever—"

"I know," Palo said, patting him on the shoulder. "I know, but it had to be done. And this isn't over. We'll figure a way."

Close by, Alfer quipped, "We're all going to starve in the winter anyway."

Moctu scowled, but worried that he was right.

"So, what do we do now?" Lehoy asked.

"We *hunt*," Alfer said adamantly. "And we hope we can kill a lot. Need meat. Need it bad."

"We should move," Nabu said worriedly. "Getting light."

"He's right," Lehoy said. "They could change their minds when they see how few we are."

"You guys drift back there," Moctu said pointing to the southwest. Stay to the south in the unburned brush, but stay close. I'm going to wait for Nuri."

The sun was fully past the eastern hills before the Lion warriors released Nuri. Moctu's gloom lifted a little as he saw her begin to walk his way. She appeared to have no injuries, but tears streamed down her cheeks, and she looked as stricken as he felt. Elka, too, seemed uninjured except for a slight limp as she began to slowly pull a fully laden travois in the other direction.

An injured Lion warrior—the one Lehoy had called Gelari—shook his fist at Moctu as he pulled a skin off the travois load, exposing a dead body. Two other wounded Lion warriors were being assisted as they walked.

"Crap, did I give up too early?" Moctu wondered. "Could we have beaten them? They only had…what, eight healthy warriors to our five."

He shook his head, remembering that, not only did they have more warriors, they had knives to the throats of both Elka and Nuri.

Nuri was sobbing as she ran into Moctu's arms. "I'm so sorry…I tried. We almost… Oh, Spirits, what do we do now?"

Moctu's jaw set as he held his weeping mate.

"You're going to join up with Palo and the others. I'm going to go to the Lion People shelter, and see if I can trade…give them anything they want to get Elka back."

"I'm going with you," Nuri said quickly.

"No, you're not. It's not going to work, but I have to try. May get myself killed—but no sense in both of us going. And Devu…he's always wanted you. I would never trust him with your life."

"I really don't think it's safe for you to go either," Nuri said. There was a forlorn silence while they both contemplated the situation. When Moctu began collecting his gear, Nuri asked, "How about Hawk and the Krog? Can they help us?"

"That's an interesting idea," Moctu said, considering.

"Elka's half Krog. They'll want her back and safe almost as much as we do," Nuri added.

"I can't see how that'd help," Moctu said finally. "Just get a lot of them killed. They don't have bows or atlatls. Like the Shiv, they're terrific at camouflage, so some might survive. But I don't think it's a good idea." Moctu hugged her hard. "I'm just glad you're safe. "Now it's best for you to help Palo and the others have a good hunt. That's what'll help our people more than anything."

"So, you *are* going to go…to the Lion shelter?"

Moctu shrugged. "Have to. It's the only thing left to try."

Chapter Fifty-Six

Her head down, Elka strained to get the travois up the other side of the charred, boulder-strewn wash they'd just crossed. The travois was heaped high with sleeping furs, tools and the dead body, and it was much heavier than before. Despite the crisp breeze and the overcast skies, Elka's tunic was drenched in sweat, and tangled strands of her hair were matted to the sides of her face.

Since the attempted rescue force had been driven off, Elka had barely slept, and her eyes were red with dark circles below. Although the daytime temperatures were relatively pleasant, nights were extremely cold, and Elka had been given only a small, thin sleeping fur. Even with Misha next to her, nights were miserable.

But it was her future that kept her from sleep. As they neared the Lion People shelter, her impending doom—the upcoming sacrifice—weighed evermore heavily on her. There were moments when, even at rest, Elka found it difficult to breathe, and several times a day she fought off the same thought: Let's just get this over with.

More and more, Elka found herself mumbling in a jumbled mix of Krog and Nerean—words that were unintelligible to anyone near her.

"We came *so* close. Almost escaped. But now..." She pursed her lips and shook her head. "They've given up on me. Every step I take gets me closer to my end. Might be different if I'd had a full life. If I'd been loved...or had this baby." Elka took a hand off the travois to feel her stomach. "Never going to meet you, little one. Never going to know if you're my son or daughter."

Misha limped beside her and seemed no better than the day before. But she was no worse either, and at least she was keeping up. Needing momentum to make it up slopes, Misha lumbered awkwardly, but quickly up the far side of the ravine and settled down at the top to rest and wait for Elka.

A tall, young warrior with two, long braids of shiny black hair tied with red ochre-stained, leather strips pointed and said jokingly, "Hey, look, there's a wolf up there we can kill."

Exhausted and hungry, Elka paused and closed her eyes, taking several deep breaths to control the anger that flared inside her. She knew she was being goaded, and she eyed the young man briefly, then went back to hauling her load.

The warrior frowned as she looked away without commenting. Wanting a reaction from Elka, he followed up with, "It's a big one. I can get him." He started to string his bow, and Elka paused again, losing some of her progress as the travois slipped backward a few feet downslope.

"Louter," Naramsin said sternly, warning the youth off.

"Oh, he's just having fun," Gelari called from farther away.

"They're not much to eat, but their skins are okay," the young man went on loudly, smirking as he ostentatiously pulled an arrow from the quiver that he wore on his side. He was fitting the arrow to the string when Elka's irritation exploded, and she dropped the travois and charged the young warrior.

The travois swiveled on the slope and spilled much of its load, including the dead body. The snickering lad was enjoying the attention of the other warriors when he heard the disturbance behind him, and he saw the look of alarm on the faces of his onlookers. He turned just as Elka hit him, her head down and her powerful right shoulder taking him in the midsection.

Virtually every mouth fell open as the bow and arrow went flying, and the boy was lifted several feet into the air and flew for an astonishingly long moment before hitting the ground and rolling. He ended up in a heap three paces downhill where he lay still.

Misha saw the fight and limped downhill to protect Elka.

Still angry, Elka turned to face the onlooking warriors, her face red and her teeth bared.

Several drew their bows and seemed ready to shoot.

"She's…she's part wolf herself," a warrior exclaimed, pointing at Elka worriedly, as Misha drew up beside her and faced the men, letting out a deep, threatening growl.

"How could she do that?" another asked. "She's a young girl, but she threw him so—"

"Alright, bows down. Now," Naramsin commanded loudly. "We gave our word she wouldn't be harmed on this journey."

"But…"

"Naramsin's right," Gelari said reluctantly. "Bow's down. She'll get her punishment soon enough."

"She *is* a wolf," one of the warriors grumbled as the bows came down. "That's how she can talk with them. It's creepy."

The men helped the young man up, and after a venomous glance at Elka, he limped away.

Elka righted the travois, and as she reloaded it, Naramsin came to help.

"That was…unfortunate," he said. "Louter is Rolf's only son. His only living son anyway. His first was killed by your…uh…by the Pale Ones. And Rolf is our leader." Under his breath, Naramsin muttered, "Although sometimes it seems like Devu is…"

Naramsin walked alongside Elka as she crested the hill with her load. Her dark mood was deep and encompassing, but in most ways, it was totally different from the sadness times she'd had in the past. Those made her cry and not want to eat. She could never figure out what was wrong. Now, she knew for sure what was wrong. She and her baby had no future. The misery that enveloped her was edged with a wall of anger, and she *desperately* wanted to eat.

Naramsin eyed her as she continued in stony silence, her lips pressed tightly together.

"Have you…" he started, then thought better about speaking. But his urge to bring a speck of brightness to her gloomy world overcame him, and he said, "Have you noticed the onlooker who follows us?"

Elka stopped and followed his gaze to the rear. Her eyes widened as she saw him. On a low, fire-darkened hill, far behind them, was Moctu.

“Father,” Elka exclaimed as a sudden warmth radiated through her body. “So, they haven’t given up on me,” she murmured.

Her next steps were suffused by a light-hearted glow of hope tempered only a little by worry for her father’s safety. He’s clever and knows how to take care of himself, she thought, nodding confidently.

The feeling was such a change from the past couple of days that the travois seemed lighter, and there was the hint of a spring in her step. That lasted until she crested the next low ridge and saw the Lion People shelter in the distance.

Chapter Fifty-Seven

"Ah, we have Sleeps With Wolves," Devu said, smiling and rubbing his boney hands together. "This changes everything."

"Yes, but Beotuk was killed and we have four others seriously wounded," Naramsin said as he and Gelari reported to the shaman.

"A hard trade, but worth it," Devu said waving them off. "Definitely worth it. Bring her to me immediately."

Naramsin frowned and studied the gangly, hollow-cheeked shaman for a long moment, noting that Devu looked even grayer and more cadaverous than usual. His hair was unkempt, both dirty and disheveled, and his beard was clotted with dark material. There was a darkness within the man, but right now his eyes were glowing.

"What about the wolf?" Naramsin asked finally. "She follows Elk…uh, she follows Sleeps With Wolves everywhere."

"Kill it, stake it down, whatever—I don't care." Devu said turning away almost bouncing as he moved. Then he spun back and held up a hand. "Wait. The wolf…" He paused, thinking, his hand pulling on his scraggly beard. "No. Cage the wolf. It'll be useful in the Ceremony of Sacrifice, and it may help us control Sleeps With Wolves better. Oh Spirits, this is good news," he said loudly, his eyes glowing. "We've been after this for so long. It's really going to help. Now, go get her," he said, waving the men off.

A skin was thrown over Misha's head, and the wolf was tied down while men and women worked on building a strong cage. Elka called to the yelping wolf, attempting to calm her, but Naramsin came to lead her away.

Gelari had tied her hands behind her, not wanting another demonstration of how strong this young girl was. Elka held her chin up defiantly as she was led past people who yelled insults at her, sometimes spitting in her direction.

"Been at war with these people for years," Elka murmured to herself. "And they blame me. No wonder they hate me."

She tried to keep that thought as people, mostly women, threw not only insults, but small rocks and refuse at her.

"Bitch!"

"Hyena dung!"

A slender, pretty woman with a dozen long braids threw a stick, hitting Elka's knee.

"You left my mate to die," she shouted. "You're evil."

"That's Mia," Naramsin whispered. "Pele's mate."

At the mention of that name, Elka froze for a moment, then looked back at the woman she'd just passed. Even though she was scowling, Mia was indeed pretty with beautifully styled hair, large eyes and a shapely body.

Noticing Elka's keen interest, Naramsin added, "Pele's still out on a hunt. Mia's one of the daughters of Devu. At least he purified her mother when—"

His words were cut off as a heavyset woman yelled, "You bitch—you killed my son." Then she rushed at Elka waving a stout pole. She delivered a hard blow to Elka's shoulder before being restrained by Naramsin. The woman was sobbing as Naramsin directed her back to a hearth.

Her hands still bound behind, Elka grimaced as she flexed her shoulder and neck, trying to ease the pain and determine if anything was broken.

"Sorry about that," Naramsin said as he returned. "Some of the people here have lost loved ones during the years of fighting. That's Beotuk's mother. He's the…it's his body you were carrying." Naramsin shook his head sadly remembering the boy. Pointing ahead, he said. "Let's get you to the shaman."

Elka clenched her fist as her head spun with pain and too many thoughts knocking into one another. But the word 'shaman' sent a chill down her spine. That would be Devu, the powerful sorcerer who had wanted her dead—or alive to sacrifice—for years.

"I've heard his name. He's the one who thinks I'm sickening, an offense to the Spirits. Seems like most of your people agree."

"Yeah, over the years Devu has convinced just about everybody," Naramsin said.

Elka cocked her head and looked at the man's scarred face, because something in his words or tone made it sound like he *didn't* agree with Devu. She couldn't see Naramsin's good eye, and the dead gray one revealed no emotion.

"Here it is," Naramsin said, nodding at a large hearth that had an assortment of skulls and large rocks lying just outside the entrance. The hearth opening was covered by a large bear skin.

"Devu, I'm here with Sleeps With Wolves," Naramsin announced loudly.

Elka's heart beat faster as she studied the tall, gaunt, scruffy man who pulled the skin aside. He examined her with an excited curiosity.

"Sleeps With Wolves," Devu said with a slow smile that built into a crooked grin. Placing a boney hand on her shoulder, he ushered her inside. As his hand lingered on her shoulder, he repeated, "Come in, come in, come in."

To Naramsin he said, "It's best you stay outside." Then Devu pulled the bear skin closed.

Chapter Fifty-Eight

"What the...?" Moctu muttered. Ahead was another large area where all the vegetation had been burned. "Must have been a huge fire. Won't be able to follow the Lion warriors without being seen now. Palo's right—I've never seen such a big stretch of burned country."

His shoulders slumped as he surveyed the area ahead and to the north. All of it was desolate and blackened except for some shady, white stretches still covered by the frost of the night before. Nothing remained except a few low, charred stumps. Nothing moved. There was no noise—no insect buzz, no bird song—nothing but a cold breeze that smelled of soot. It was eerie, and the barren sight deepened his bleak mood.

After a long look at the empty, dark land, Moctu shook his head and turned his gaze to the south.

"If I move south, I can keep to the unburned brush. Stay out of sight and maybe find some rabbits or quail. I'm starving," he murmured.

The morning sun had crested the highest eastern hills, and it was partially veiled by billowing, white-to-light-gray clouds whose backlit edges glowed silver. As Moctu moved quickly through the brush, he was more aware of the multitude of different shades of green and brown vegetation. In contrast to the vacant, desolate land to the north, it seemed to calm and settle him.

The sun was high, and the shadows had just turned when he saw it. On a broad, grassy hill a rhino was browsing on the low sedge. It was a big one, and Moctu immediately noted the two arrows that protruded from its left side. The animal was alert and healthy, obviously unperturbed by the arrows. The weathered, gray look of the shafts and the lack of any feathers suggested that they'd been there a long time.

Moctu began to pull an arrow from his side quiver, but he released it and snorted sardonically.

"No way I could bring that down myself. Isn't that the way it goes?" he said, smiling wryly. "Palo and Alfer and the rest are hunting and probably finding nothing, and I'm here a day away from the Lion shelter, and there's a rhino that would feed our people for a moon." This is a sign from the Spirits, but what are they trying to tell me?"

Moctu rubbed his short beard thoughtfully. Still, he thought, if I could take it down, it would give me something to trade with the Lion People for Elka. Maybe they're like us and don't have enough food stored for winter.

Another look at the rhino, and he dismissed the thought.

"It's huge, and I only have six good arrows. In the past, it's taken…sometimes as much as four hands of spears and darts to bring one down. I'd have to hit it in the eye or some other…" He expelled an exasperated breath. "No, I don't think there's any other place I could hit it to bring it down myself. And even that might not be enough."

The irony of the situation overwhelmed him, and Moctu sat down dejectedly and just stared at the beast. For the first time, the rhino noticed him, and it stopped chewing and eyed him for a few heartbeats before it went back to foraging.

"What *am* I going to offer Devu and Rolf in trade for Elka?" Moctu wondered. "We don't have any spare food—don't even have enough for ourselves. We have some bows and the two boats. Maybe more of our hunting territory?" He shook his head and watched the rhino, sure that the Spirits were either telling him something or maybe just teasing him.

"It's times like this that make you wonder," Moctu mused, his eyes now looking vacantly at a stand of grass a few feet away. "There've been times that I know there are no Spirits…and there are times when I'm certain of them. Sometimes, it's like you can sense the Spirits. They're all around you. And they're all there. Other times…"

He shrugged. "Other times like now..." He frowned and shook his head. "So, which is it? If you're out there, I need help."

Chapter Fifty-Nine

INSIDE DEVU'S HEARTH IT was stuffy and dim. As her eyes adjusted, Elka could see the space was cluttered with skulls and baskets of bones. She was stunned to see that some of the rocks along the wall—sizable boulders—had bones and seashells sticking out of them.

"You're different than I expected," Devu said. "You look almost like a real...well, almost like one of us."

Elka wrinkled her face at the comment, but didn't reply. Careful not to alert Devu, she continued to pull hard at the binding that bound her wrists behind her. Her skin was rubbed raw from yanking against the rough, leather straps. But despite her straining, the straps wouldn't yield.

"I saw you when you were a baby, but now you're..." Devu said a little breathlessly. "I see that you really do have blue eyes. I thought that was probably just a rumor."

Devu cleared his throat and placed a hand on Elka's shoulder. In a softer voice he said, "You know, I have the power to delay or even stop your sacrifice which is planned for this upcoming full moon. I could easily delay it into the winter when the daylight is short and the darkness is at its longest."

Elka's heart beat faster, and she discontinued her struggle with the bindings. She turned her head to face Devu, one eyebrow raised.

"You don't speak, do you?" Devu asked. "I wonder if that's—"

"I can speak," Elka interrupted.

"Oh, good. Well, as I was saying, I can delay the sacrifice depending on how...*cooperative you are*." The shaman's voice deepened on the last few words, and Elka's head recoiled a little.

Remembering Nuri's account of her time with Devu, Elka had a good idea of what the shaman would say next.

In a husky voice, Devu said, "I can do my best to…purify you, so that, perhaps, a sacrifice isn't needed."

The urge to headbutt the shaman almost overwhelmed Elka. She knew she could break the man's thin, boney nose and maybe even kill him. But instead, she bit her lip and looked away.

Have my baby to protect, she thought, blowing out a breath. And Moctu's out there somewhere. I need time—need to stay alive.

"I'm pretty sure I know what you mean," Elka replied frowning. "Give me some time to think on that."

There was a gleam in Devu's eyes as he leered at her. He smiled and said, "I can give you a few days, but keep in mind, with every day that goes by, the moon gets fuller."

Chapter Sixty

"Well, at least we won't starve for another moon," Alfer said sourly, as the group viewed the downed horse. They were all breathing heavily from the chase of the wounded animal, which had taken all morning. Palo, who had scored the first hit on the young stallion, moved closer and fired another arrow into its chest to ensure it was dead.

He knelt carefully beside its head and placed his hand on the neck. In a low voice he said, "You're a fine beast, and you battled hard. We're sorry to end your life, but our people need to eat. Thank you for feeding us."

"Yeah, yeah," Alfer said dismissively. "We'll need a hand or two more of these to survive the winter. And this is the only big animal we've seen. So much burned-out country."

"Alfer's right," Palo said, standing up. "We need more meat, so some of us should keep hunting. But this meat needs to get back to Etseh. Who's going to do that?"

"I'll go. I need to get back to Etseh and make sure Avi's okay," Lehoy said.

"You can't pull all this meat by yourself," Palo said. "Who else wants to help Lehoy?"

Most of the mens' eyes shifted to Nuri, but she looked away, conflicted by the multitude of things she needed to be doing. Most of all, she wanted to be with Moctu, helping him rescue Elka.

Someone probably should go north to alert Rah and the rest of the Krog that Elka was in trouble. After all, she was one of their own—half theirs anyway. Like Moctu, Nuri knew the Pale Ones couldn't help, but at least they should know. And nobody would be going to their shelter if not her. And then, of course, Nuri needed and wanted to be with her best friend, Avi, helping her with a difficult pregnancy.

"Okay, I'll go," Nuri volunteered finally. I want to check on Avi too. We'll make two travois, and I can pull one. And as long as Avi's okay, I'm going to go find Moctu and see if I can help him get Elka back."

The group rested while they grilled and feasted on the horse's liver and heart. It was delicious—the first big meal they'd had in more than a quarter-moon. At mid-afternoon, Palo and Nabu jostled Alfer awake, and the three of them moved northwest, first crossing another desolate, burned area to get to the forested foothills.

While Nuri cleaned and butchered the horse, Lehoy built two travois—a big one for him and a smaller one for Nuri to pull. It was dusk before they loaded the travois and finished their other duties, so they feasted again and built a big fire to keep wolves away. Nuri took the first watch while Lehoy slept. The half-moon was high in the sky when she heard the first howls and barking. Wolves—and a lot of them.

There was no need to wake Lehoy. He was up and stringing his bow after the first few nearby howls. Nuri quickly gathered more firewood, glad that they hadn't stopped in one of the burned-out areas. She stoked the fire then strung her own bow. Back-to-back, on the side of the fire with the travois they awaited the wolf pack.

"There's a lot of 'em," Lehoy said after swallowing hard. "How many arrows you got?"

"Six, plus one without a good point. How about you?"

"Six, but two don't have fletching. Plus, I've got the spear on top of the travois there."

There was a long silence broken only by the howls and barking of the approaching wolf pack.

"Sounds like…" Lehoy cleared his throat. "Sounds like we might not have enough arrows."

"Yeah, we can't afford to miss," Nuri said resolutely. "Now I know how Moctu felt all those years ago." Both she and Lehoy had heard the tale many times around the evening fire about the four boys fighting off a wolf pack that wanted their fresh meat. "At least we have bows, not atlatls."

"There's one," Lehoy said, quickly raising and pulling his bow, but not firing.

The first wolf was soon joined by others, and in the dim light, it looked to Nuri like a swarming, circling sea of the huge, fury beasts. They were yipping and barking excitedly, eager and impatient, but for the time being, hesitant to approach the fire.

"Whew, there's so many," Nuri whispered, her mouth dry.

"And they're hungry," Lehoy added. "Not as much game for them because of all the burned-out areas. They're starving—desperate."

The night sky was clear, and a cold wind was blowing from the north. It would frost again tonight. But Nuri felt sweat trickle down the side of her face. She took her hand from the bow string and touched her talisman, a red hawk feather she wore at her neck between two, small ivory rings.

Closing her eyes briefly, she quietly said, "Spirits, please be with us. Please help us. *Please*."

"There's so many of them, it's almost like we can't miss," Lehoy said, and Nuri nodded, having thought the same thing. But neither of them fired an arrow, knowing they had to make every one count.

A few wolves began to make lunging dashes toward the travois, but the clever beasts bolted away before they could be targeted. The other wolves, emboldened by the more courageous or desperate ones, edged closer.

Lehoy let loose the first arrow and scored a hit, causing a wolf to yelp loudly and careen away.

Likewise, Nuri fired at a large wolf that slowed while circling the camp, hitting its hindquarters.

Buoyed by their initial success, each of them shot twice more in rapid succession. Lehoy's third shot grazed an animal, but neither made a solid hit.

"Whoa—gotta slow down. That's half our arrows," Lehoy called loudly. "I can't believe how quick these guys are."

Yeah, it's like they sense when we're going to shoot," Nuri replied. "And when one of them makes a run at you, it's hard not to fire."

Unconsciously, Nuri and Lehoy had moved closer to one another, and they were almost back-to-back.

"Should we let them have some of the meat? Maybe throw some out there?" Nuri suggested in a hushed voice. Then, with a nervous laugh she said, "Don't know why I'm whispering—it's not like they can understand."

"Yeah, but it's eerie how much they can sense what we're going to do," Lehoy said quietly. "But I don't think we should throw meat to them. Probably just excite them more."

"I'm going to carry more arrows from now on—that's for sure," Nuri said. "If we get out of here," she added under her breath.

"Me too," Lehoy agreed, just as he fired an arrow at a giant, charging beast. The arrow took the wolf in the chest, staggering it. It was a fatal wound, but the brute yelped and managed to limp away.

Farther away in the darkness, they heard what sounded like a fight between wolves, and Nuri remembered that hungry wolves will sometimes attack and eat wounded members of their own.

Although still circling the camp, the wolves began to move deeper into the darkness, and they made far fewer charges into the firelight.

"Maybe you hit the leader. Maybe that big one was leader," Nuri said shakily.

"They're backing away," Lehoy said hopefully. "Can only see a few of them now."

And then there was stillness.

Nuri's eyes widened as she listened harder. The edges of her mouth curled upward as she began to let herself believe the pack was gone, and she would survive.

After a long silence, Lehoy blew out a breath. "They're gone."

They turned to each other and embraced, bouncing with joy. It was a wonderful feeling, and even though she was still trembling, Nuri let out a whoop of relief. That's when Lehoy kissed her.

Chapter Sixty-One

ELKA SURVEYED HER COLD, new living space, a small enclosure with no hearth for a fire and the entrance skin pulled to the side so it didn't cover the opening. It was late afternoon and already a frigid wind blew from the north. The guard outside wielded a wicked-looking spear with a long, red, sickle-shaped point. He was tall and lean, and his broad shoulders suggested he would be a formidable adversary in a fight. Even though he wore a thick fur and a bearskin hood over his head and neck, he continually paced and muttered complaints about the cold.

Elka huddled in the far corner of the enclosure where the gusts of icy wind didn't always reach. The guard had earlier wordlessly taken her moccasins, which Elka supposed was to make her colder and even more miserable. It would also keep her from running away even if she managed to get free. The guard then untied her hands from behind her and retied them in front. He'd pointed to a partially hollowed-out, short section of a tree trunk that was to be where she relieved herself. It was covered with flies and smelled awful the one time she used it.

Misha's heartbreaking howls and whining still came often from far off to her right. Yesterday, it had been nearly constant. Now, it was softer and not as frequent. It tore at Elka's heart not to be able to go comfort her, and Elka felt a sadness time coming. Knowing she couldn't afford to be overwhelmed by those feelings, Elka forced herself to focus on the anger she felt instead of the hopelessness, and the feeling passed.

A long moment later, her thoughts drifted to introspective curiosity. She was surprised to find her face still frozen in a scowl, and that quickly faded as she stared at the ground perplexed.

"I haven't had any…" she murmured, putting her hand to her small chin. "None of the sadness times have come for… How long *has* it been? None since…at least when we were so hungry and hurt, and the wolves attacked. Is that right?" She made a small, disbelieving shake of her head. "All sorts of problems and troubles—plenty to be sad about. Yet…nothing." She gave a sardonic half chuckle and smiled quizzically. "Now, when I have every reason to be sad…" This time, she gave an actual snort of laughter at the irony of it.

"Okay then. What can I do now?"

Whenever the guard wasn't looking, Elka began sawing the bindings on her wrists against a small ridge of sandstone on the rock wall. The small exertion helped take the edge off her chill and keep her shivering to a minimum.

"Making progress," she murmured to herself as she examined the bindings in the murky light. "But what do I do when I'm free?"

Besides the guard, other inhabitants, mostly women, passed by regularly. Many paused, peering into the dim enclosure, trying to see the girl they'd been convinced was the root of all of their problems. Most of those who paused reviled her with insults, often spitting into the entrance. The guard made no attempt to discourage this behavior, but he did keep them from entering the enclosure or throwing things into it.

"Elka. Elka, you can…hear me?"

Elka's eyes widened, and she went stiff as she recognized the voice. It was Pele. He was standing with the setting sun behind him, so all she could see was a silhouette. Her heartbeat quickened, and she was filled with a confusing mix of emotions—everything from hate to yearning. But the hate quickly settled in as Elka recalled the murder of her brother and Ender, and the injuries inflicted on her father, herself and Misha.

"So, you got what you wanted," Elka called out to him, her lip curling. "I'm here, and they'll be sacrificing me soon."

Pele entered the enclosure, but was blocked by the guard.

"Elka, I—"

"Just go," Elka said loudly. "Just get away from me."

Behind Pele, there was a new, smaller silhouette that touched, then pulled on his shoulder.

"Come love—you shouldn't be here. Come."

"So that's Mia," Elka muttered to herself.

Pele hesitated, turned away, then back. "I come again tomorrow."

"That's not a good idea," Mia said. "Now come," she said pulling harder on his arm.

"Yeah, that's not a good idea," Elka said sourly in a loud voice. "Listen to Mia, *your mate*."

Pele blew out an audible breath, then slowly allowed himself to be led away.

Elka opened her mouth to call him a murderer, but thought better of it. She looked away and put a bound hand to her forehead, feeling the throbbing pulse on her temple. There was a tightness in her chest that she'd never experienced before, and she was shocked to feel the wetness on her cheeks.

"Why am I crying?" she muttered angrily. "That…" She shook her head but the tears were flowing freely now. "That murdering, filthy hyena."

Chapter Sixty-Two

THE KISS TOOK NURI by surprise. Before she could react, Lehoy's tongue was in her mouth and his hand cupped her bottom, pushing it into his growing hardness. His embrace was strong, but she pushed a hand's breath away, turning her head to the side.

"No," she said firmly but calmly.

It was dangerous for women to shut down and frustrate a man too abruptly. Especially a powerful, self-assured man like Lehoy.

Nuri remembered her mother's admonition, "Men are often brutal because they have unstable hearts. They're friendly until they want you, then they're vicious jackals when they don't get their way. You can soften their hearts with calm words that don't embarrass them. And it's best not to be alone with a man for too long, unless he's your mate. It's not fair, but that's the way it is."

Nuri knew of a lot of women who'd suffered from men with uncontrolled desires. She'd already experienced it too. Thoughts of Jabil and Shaka-Nu sprang to mind.

"Avi and Moctu wouldn't like this," Nuri said, trying, but failing to push farther away.

"They don't have to know," Lehoy said in a husky voice, pulling her closer.

"No. It's not going to happen," Nuri said more forcefully. Then she softened her words by saying, "Avi's my best friend. I'm flattered by your interest, but this *can't happen*."

Her words finally had an effect, and Lehoy relented, allowing Nuri to pull away.

The tall, well-built man blew out a breath and looked down at the ground shaking his head. "Don't know what came over me," he said sheepishly. "I…" He shook his head again. "Please don't tell

Avi about this. Or Moctu. He's my best friend." He looked into Nuri's eyes and said, "*Please*."

Nuri reflected on that for just a moment. Saying something would certainly ruin two friendships. She nodded at Lehoy and said, "No one needs to know about this as long as it never happens again."

"You mean that?" Lehoy asked hopefully. "You won't say anything about this?"

"About what?" Nuri said smiling broadly. "Now let's stoke the fire with more wood so those beasts don't come back."

Chapter Sixty-Three

"Crap, how is this ever going to work?" Moctu said, frowning as a group of five warriors began running toward him with their bows drawn. "Every time I wave a branch for a truce, they come running to kill me."

With that he turned and ran, melting into the brush south of the Lion People shelter. This area was higher than the lowlands he'd come through earlier, and it had lots of rolling, low hills.

Moving quickly but quietly, Moctu shook his head in frustration. This was the second time he'd tried for a parley, waving a large, green oak branch above his head only to have a number of hostile Lion warriors pursue him. They were obviously set on bringing him back to Devu with enough arrows in him to look like a porcupine.

Moctu's only advantage was that he could move rapidly through the brush and young forest in any direction, as long as he avoided the burned-out areas. His pursuers not only had to track him, they had to be careful not to be picked off by a hidden bowman.

"Once I get there... *if* I get there, what do I offer in exchange for Elka?" Moctu muttered.

It was a subject that never left his mind. He just knew he had to do this. Every fiber, every bone in his body told him this was obligatory—an inviolate responsibility of fatherhood. And then, there was the promise he'd made to Effie. But every part of him also screamed that this effort was hopeless and doomed to failure. No matter, face-to-face with Devu, he would offer just about anything to get Elka back.

The sound of a twig snapping brought his thoughts back to the present. It was off to his right about a half bow shot away. Moctu instinctively began nocking an arrow, but thought better of it.

“I won’t have any chance for success if I kill a bunch of their warriors,” he murmured.

With that, Moctu turned and silently moved in the opposite direction. Still out of sight, moving north in low brush, he crested a knoll and saw that the next hills beyond were blackened by fire and barren of cover.

“So that’s what you guys are trying to do,” he muttered. “Drive me into the open.”

He surveyed the surrounding hills, most of which were scorched, but many of the shallow valleys still had vegetation. He remembered his father, Tabar telling him that forest fires travel uphill faster than down.

“Just like a campfire, the heat moves upward, drying out the higher vegetation which then catches fire easier,” Tabar had counseled. “Stay as far away as possible, but if a fire comes your way, it’s usually best to move in the opposite direction along the lower, wetter routes. But, no matter what, they’re dangerous because they’re so unpredictable.”

The sight was depressing, but bizarre, looking to Moctu—as if a huge black hand had swatted the earth, leaving the imprint of charred fingers along all the ridges.

“Can’t stay here,” Moctu said after hearing a shout from a pursuer. “They’re too close.”

Careful not to leave tracks, he headed down into the vegetated valley where he took cover in a small grove of oak trees.

“It’ll be dark soon enough,” Moctu murmured, smiling as a plan came to him about how he might break the stalemate. “I hope they have another fire tonight.”

The warriors trailing him got no closer that afternoon, and soon the sun dipped below the hills. With the twilight came a wintery breeze that dampened the night songs of the frogs, crickets and other evening noisemakers.

“Yeah, they’ll have a fire tonight,” Moctu said, shivering but smiling. “I’m so hungry. I might even get something to eat.” He wrapped his small fur tighter around his shoulders and settled down to wait.

It was a clear night, and the moon rose high casting its faint light over the brush and trees. Moctu frowned, noting that the moon's left side was bulging, now well past the halfway point. He knew that most important sacrifices happened at the full moon, and his chances for saving Elka were dwindling down.

"Time to get this done," Moctu said, restringing his bow. Moving up the far side of the valley, he broke out of the brush and into the dead, burned-out finger that ran along the crest of the hill. In the open, it was easy to spot the large campfire of his pursuers downwind, not far to the southwest.

There was just enough light from the moon and stars for Moctu to move quickly along the edge of the vegetation without making noise. As he neared the campfire he slowed to determine where the guard was, or whether there might be two. Two would be a real problem.

The breeze blowing from the northeast was colder now, and Moctu found himself shivering.

Is that from the cold or my nervousness? he wondered.

To offset the cold, the Lion warriors had built their fire high, and it gave off enough light for Moctu to see the campsite well.

"There are four sleepers. So, where the other guy?"

Moments later he saw movement on the south side of the camp.

"There you are."

Moctu nocked an arrow and ensured his others were easily at hand. Stealthily, he moved to the south toward the camp guard. He stopped once to pocket a fist-sized cobble that he felt through his moccasins.

"Just the right size,' he murmured smiling.

Moving painstakingly slow through the trees and brush, he soundlessly neared the guard. The young man's bow was strung, but he had no arrow nocked. He leaned against a tree and appeared to be half asleep.

Moctu pulled the rock from his pocket and thought through the logistics of holding his bow while hitting the guard, then easing him to the ground without waking the others.

Hit him hard, but don't kill him, Moctu cautioned himself.

It all went pretty much to plan. After he struck the guard, the young man fell against Moctu's chest, and he was able to ease him to the ground. The only noise had been the guard's bow hitting the ground. Scanning the camp, Moctu was relieved to see that the other warriors were still asleep.

Moctu unstrung the guard's bow and pocketed the sinew, then checked his own arrows again making sure he could easily pull them from his quiver. Then he took a position beside a good-sized pine tree whose dead, lower branches had been recently broken off for kindling.

"Okay," he murmured. "Got a good field of fire, I can see everyone of them, and my arrows are all ready." He nodded to himself, then pulled his bow back.

"Alright," he said. "This is how it's going to go."

Three of the men immediately shot up to seated positions, still wrapped in their furs. The other warrior was on his knees and had grabbed his unstrung bow.

"Stay down and don't touch your bows if you want to live," Moctu commanded, his tone thundering and menacing.

The sitting warriors, still groggy from sleep, froze, and the kneeling man released his grip on his bow.

"Who leads here?" Moctu roared.

The seated man nearest him, a brawny, older warrior nodded and said, "I lead this group."

"What's your name, leader?"

"I am Shintow," the man said somewhat sheepishly.

"Well, Shintow," Moctu said loudly. "My arrows are ready, and your bows are unstrung. Your guard is down but not dead. I can kill you all, but I won't if I don't have to. If you give me your word of honor that I won't be harmed, and you'll take me to Devu and Rolf, if he's still your leader, then I'll put my bow down and come with you in peace. Do I have your word?"

Shintow was relieved, and answered immediately. "Yes, I give you my word."

"I will not be harmed, and I'll be treated with respect, as I'm coming in peace for talks?" Moctu questioned him further.

“Yes, you have my word,” Shintow reiterated.

“With all your men as witnesses to your word, and with the Spirits watching and hearing, I’m therefore putting my bow down,” Moctu said, gradually easing the string on his bow and placing his arrow in his quiver.

The warriors were immediately up and the camp was in an uproar, but Shintow quickly moved between his men and Moctu, calming them.

“Okay,” Moctu murmured to himself with a smile. “No turning back now.”

Chapter Sixty-Four

Nuri stared at the western horizon, a blaze of golden sunbeams streaming between reddish-gray clouds and through breaks in the dark blue hills. Sprays of light shown down through the gaps in the clouds all the way to the ground, seeming walkways to the heavens. The sight was beautiful, but it did nothing to ease the gloom in her heart.

Rather than the delight that hunters usually received upon bring in a large portion of game back to Etseh, Nuri and Lehoy had been met with a tense foreboding. They'd quickly been ushered to where Avi lay very pregnant, struggling to breathe, her skin a dreadful, cadaverous gray. She was sweating profusely, her light-brown hair matted to the sides of her head. A visibly shaken Jelli continually swabbed her face and neck with a cool, wet skin. The oldest person in the tribe, the wiry, little woman was usually commanding and confident. But now, she was hunched over and shaky, looking completely drained.

"I think the baby poisons her," Jelli said in a raspy voice while looking up to them with reddened eyes. "I've tried everything I can think of."

"Oh, Spirits," Nuri exclaimed, kneeling beside her best friend and taking her hand.

Standing behind her, Lehoy groaned.

"Oh, no," he moaned, putting his hand to his head. His eyes closed, and he shook his head softly.

"I've tried yarrow and willow bark," Jelli said. "And two different mushrooms. I've said all the prayers and chants," she added weakly.

"What does Nindai think?" Nuri asked.

"He's been sick, too," Jelli responded. "His leg has swelled up and he has fever. I tried to help him, but Avi needed me more, so now his mate, Sima, cares for him. That's one of the reasons I'm…" She blew out a breath. "So tired."

Nuri nodded. Avi was unresponsive, so after adjusting the dried grass under her head to make her more comfortable, Nuri left the crowded hearth to think.

"What can I do?" she murmured, staring absently at the sun setting over the hills. Her eyes welled with fluid, and with the fingers of both hands, she wiped the tears away. "Spirits, please let me think of a way to help Avi. *Please* give me the wisdom to help her."

There are so many things I need to be doing, she thought, rubbing the back of her neck. But first things first—gotta help Avi.

Nuri began pacing, and unknowingly she found herself at Nindai's hearth. Sima was tending him, wiping his face and chest with a cool, wet skin. Nuri was pleased to see Nindai's eyes were not only open, they widened upon recognizing her and they tracked her movements into the hearth.

As Nuri's eyes grew accustomed to the dim light, she was shocked to see the shaman's right leg was nearly twice the size it should be. As a boy, Nindai broke the leg, and it ended up about four finger-widths shorter than his left one. Now, it was dangerously swollen.

Forcing a smile at Nindai, Nuri said, "So, you're awake. Looks like Sima is taking good care of you."

Nindai tried to say something, but he only coughed instead. Clearing his throat, he croaked, "Yes, she's…" His eyes closed and he continued, "she's very good to me."

"I'm hoping you can give me some ideas on how to help Avi," Nuri said wrinkling her brow, then biting her lip. "She's so sick, and she's not responding to anything. And I think Jelli's close to getting sick herself. Do you remember what it was that Elka said helped her so much when she was injured? And what did Moctu use to help the Krog with the lekuk spotted disease? Maybe one of those…" Her voice trailed off.

Nindai shook his head and his eyes flickered open, then closed his again. A long moment passed, and just when Nuri was convinced that he was asleep, Nindai answered.

"Garlic," he said simply. "And nicroot. I…have some in my medicine pack," he said gesturing with a bent finger. "Nicroot's best when…" Nindai blew out a breath, and waited a moment to regain some strength. "When you use yarrow with it," he finished.

"Thank you Nindai. Thank you," Nuri said, quickly stooping to find the nicroot where Nindai had pointed. "I'm going to try these with Avi. And I hope you're all better very soon."

Sima was still focused on her mate, diligently wiping the wet skin across the shaman's face. Nuri studied the Pale One's broad shoulders a moment before the woman looked up.

"You're doing well, Sima. Please get him better, and let me know if I can help."

The Mung woman seemed to understand, and she nodded and smiled in appreciation.

Nuri excitedly left with the wad of nicroot, and she went directly to her hearth. Almost every hearth had yarrow, and she had plenty of garlic as well. In short order, she prepared the three herbal remedies, the nicroot and garlic ground up as pastes, and the yarrow as tea.

From a corner of the hearth, she retrieved and unwrapped the soft skin covering the bulbous Earth Mother figurine which was used during every birth. The powerful figurine imparted energy and good fortune to both the birthing mother and her child.

"This isn't a birth yet, but this seemed to help her last time," Nuri murmured, shrugging and pocketing the idol.

"Now we just need Avi to be awake enough to swallow these," Nuri said as she carefully walked with two large clam shells with the pastes and an oblong, carved wooden bowl with the tea.

Lehoy met her at the hearth entrance, his face drawn and ashen with worry. Nuri could see in the dim interior that the exhausted Jelli was asleep by the back wall. Before Lehoy moved aside to let Nuri into the hearth, he whispered, "This is my fault, isn't it?"

When Nuri looked confused, he said, "The Spirits are punishing me for…for what I did…with you."

Even as preoccupied as she was, Nuri's mind flashed to similar remarks Avi had made years before when her mate, Jondu, was mortally wounded. She frowned and shook her head.

"Take these shells…please Lehoy, before I drop one." Then Nuri shook her head more forcefully and looked sternly at the man. "No."

Nuri paused for a moment wondering if she should voice her immediate thought.

"Look, Avi said the same thing years ago when Jondu was wounded, and that was about you—*your* relationship with her. It wasn't true then, and it's not true now. Jondu was wounded by the Shiv, pure and simple. The Spirits weren't punishing Avi then, and they're not punishing you—or her—now. They're not involved in… all this. They don't…"

Nuri pursed her lips thinking of the Earth Mother figurine she carried in her pocket. Her voice faltered as she recognized the apparent inconsistency, but she recovered quickly.

"Sometimes…think of it this way—the Spirits made us, and They love us. Sometimes we get sick in spite of that. And They gave us the ability to think and remember things so that we can help ourselves in those cases."

With a thin smile, Nuri tipped her head toward the tea she held, and she pointed to the shells with paste. "That's what these are for."

Chapter Sixty-Five

THE GRAY MOTTLED SKY mirrored Moctu's mood as the group marched toward the Lion shelter. There was still frost on the ground this morning, and the cold wind from the north suggested snow was coming soon.

"Great," Moctu said to himself, scowling. "An early winter. My people are faced with starvation, and I'm here doing…" He shook his head. "What? What *am* I doing here? How can this end well?"

He shrugged and muttered, "But how can I *not* be here? She's my daughter."

True to their word, the Lion warriors had treated him well even allowing him to keep his bow.

But now, as they crested a low knoll and the Lion shelter was in sight, Moctu's heart beat faster, and he was second-guessing himself.

"Spirits, please help me get Elka out of here safely. Please give me wisdom in my talks with Devu and Rolf."

By midday, as they reached the shelter, the daylight darkened and the wind blew harder, carrying tiny flakes of snow. Moctu was shocked by how many young warriors the Lion People had, bustling about even in the poor weather.

They have so many more people than we do, he thought worriedly. Easily twice what we have. How'd that…happen? Their hunts must have been so successful over the years to produce so many babies, so many healthy young people. Have I been…a poor leader? We may have to find our kinfolk to the south. Ordu's son Marco leads them. Don't even know where they are, but the Sea People said they met them on the coast far to the south. Even with them we won't have as many as all these Lion warriors.

Moctu's breath caught as he heard the mournful howl of a nearby wolf which had to be Misha. His focus returned as he was quickly stripped of his bow and knife by the scar-faced Naramsin, and together with the burly Shintow, they ushered him into the presence of Devu, Rolf, and several others.

It was a large, broad chamber, and Devu sat in the middle, clearly acting as the head man. Rolf, the nominal Lion People leader, sat to his right but he looked distracted. On Devu's left was the heaviest woman Moctu had ever seen. He remembered Nuri telling him of a young, obese, female shaman who reminded her of the Earth Mother figurines. The Lion People must have had enormous supplies of food to feed even one person that much.

"So, the great Moctu is here to plead for the life of his cursed daughter," Devu said smiling.

"I am," Moctu said in his deepest voice. "The war between our people has gone on long enough, and too many have died or been injured. We used to live in friendship and cooperation. We exchanged our daughters to mate and raise cousins to further bind our people."

Moctu was pleased to see Rolf and the female shaman nod, so he went on.

"We hunted together and had exciting contests. We traded goods and new learnings. We fought together against the Shiv. I think all will agree that life was better then."

Moctu saw Devu's eyes flick to the right, registering Rolf's nod of agreement.

"But your daughter *is* part Shiv—the creatures we fought," Devu interjected quickly. "And she's cursed by the Spirits. Our hunts used to bring us nearly unlimited meat—more than we could use. But they've gotten worse each year, and now our people get hungry and sick."

"She's *not* part Shiv," Moctu said louder than he expected. "She's part Krog, from the group of Pale Ones that helped us *defeat* the Shiv. And your people are hungry because of all the wildfires around here. There's nothing for the game to eat."

"Those aren't..." Rolf started, then glanced at Devu. "Those aren't from wildfires. Devu has seen in his dreams..." He looked again to Devu and cleared his throat.

Moctu frowned and unconsciously cocked his head, surprised at how submissive Rolf was being.

"Devu discovered long ago…" Rolf went on. "Through his dreams that we could kill more game by steering them…by driving them with fires we set. Until three summers ago, it worked great. We went from not having enough meat to having more than we could eat. These past two summers…well, not so much."

"Spirits, that's because—" Moctu started loudly.

"It's because we're cursed by your daughter," Devu interrupted, casting a boney finger at Moctu. "We all know that prey is supplied anew each year by the Spirits. But now, they're sickened by your Sleeps With Wolves. If we sacrifice her…" He pulled on his scraggly beard thoughtfully. "Or if I can purify her, the Spirits will be revived, and our hunts will improve."

The "purification" comment was the first glimmer that Elka's life might be spared, and Moctu was heartened. But he knew Devu's meaning, and he was disgusted at the same time. But a live daughter soiled by Devu was better than a dead daughter.

"I know your fathers and grandfathers taught you the same thing that mine taught me," Moctu said in a strong voice. "Treat the earth well. Don't overhunt the herds. It seems to us when we're young, that we inherit the earth from our forefathers, but we really borrow it from our children."

Moctu quickly scanned the crowd, and he saw that his words were having an effect.

"You're wrecking…You're *ruining* the earth with your fires. There's nothing for the game to eat." As it registered on him that the fires had impacted his people, too, Moctu began to seethe.

Staring at Rolf he said, "Would you move your people into a land that was black and barren? Well, neither would our game. That's why they don't come anymore. I can't believe—"

"That's enough," Devu screamed in a high voice. "We do what the Spirits tell us, and this is what I've seen in my dreams. The earth has whispers of wisdom for those who listen, and I listen not only during the day, but at night in my dreams. I know the Spirits will be revived if we please them with a sacri—"

"It's you," Moctu thundered. "You've ruined—"

"Guards, take him away," Devu screamed even louder. Bind him, and put him in the back chamber.

"But Devu," Naramsin interjected. "We gave our word... Shintow promised that he'd be treated with respect since he came in peace. We can't—"

Shut your mouth, Scarface," Devu shouted, his face purple with rage. "Do what you're told, or you'll be in the back chamber with him. And if this piece of Nerean dung speaks anymore, gag him. We'll hear no more of his poison."

Chapter Sixty-Six

THE BINDING ON HER wrists parted and fell to the ground, but Elka quickly replaced it, arranging it to appear that her hands were still tied.

"Now what?" she wondered. "Even if I can overpower the guard, I'm not leaving without Misha, and there's lots of hearths and people between here and there."

She was glad that she still appeared to be bound, because the guard thrust his head inside the hearth and surveyed her with a smirk.

"Looks like you'll have company when they sacrifice you at the full moon. We've got your father now too." With an arrogant chuckle he left, pulling the entrance skin closed.

"Oh dung, dung, dung," Elka sighed. "They captured Father. Spirits, I hope he's alright." She shook her head dejectedly. "Wonder where they're putting him. Crap, this complicates things even more." Elka put a hand to her chin and stared vacantly at the dirt floor of the hearth for a long moment.

An idea came to her, but it was a distasteful one. She squeezed her forehead hard, trying to come up with something different, something less repulsive.

If I submit to Devu's 'purification' maybe I can get him to release Father and Misha, she thought, immediately shaking her head and frowning in disgust.

"Don't think I can do that," she muttered, even now recognizing that it was the only realistic course of action. "Oh, crap. Oh, Spirits… no, no, *no*."

Mother Nuri had told her of when she was faced with a similar choice not once, but twice. Once with the Shiv leader, Shaka-Nu, then later, with Devu. She'd avoided Devu's sexual advances first by head-butting him, then ultimately, by showing him she was a sorceress who could make fire.

"But everyone can make fire now, so that won't work," Elka muttered. "I know I can kill him, even if my hands are tied. I can break his scrawny nose with a headbutt. Then more headbutting or stomping on his face and neck when he's down."

She smiled at the grizzly mental image, but her smile faded as she realized that would end up with Moctu, Misha and her all dead.

"But if I go through with this, Devu's seed will affect my baby," Elka worried.

All women knew that a baby shared characteristics of each of the men who contributed seed during pregnancy. That's why Nuri's pregnant mother had been encouraged by her mate to have sex with Samar the toolmaker several times to impart his creativity and craftmanship into the baby—and it had worked. Nuri was extremely creative.

"But my baby's already cursed having Pele's rotten seed. Now we're going to worsen it with Devu's seed?"

Elka frowned, searching for another option—any other option. But every alternative she imagined ended in the deaths of Moctu, Misha and her—and her baby.

"Tell Devu that I need to see him," Elka yelled out to the guard. "Tell him I agree to his…his suggestion. He'll know what I mean."

With a perplexed look, the guard stuck his head inside the hearth entrance, and Elka repeated her comments.

The guard withdrew, but he didn't leave his post, fearing that this might be a trick by Elka to escape. He yelled for support—someone to take over his guard duty so that he could deliver an important message to Devu. Very shortly there was commotion outside and Elka heard a man say, "I take over. I guard now."

Elka brow wrinkled as she immediately recognized Pele's voice. She struggled to catch her breath as she recoiled to the back of the hearth.

"Thanks, I'll be back soon," her guard called, already some distance to the left of her hearth on his way to see Devu.

Elka's lip curled, and she readied herself to attack Pele if he came into the hearth.

"I watch," Pele replied reassuringly.

It was only moments before Pele stuck his head inside the entrance, and Elka's muscles tensed, preparing to charge him.

He's big, but if I hit him just right, Elka thought.

Pele's next words stunned her.

Chapter Sixty-Seven

Nuri's potions had minimal effect. Avi was still very sick, but at least she hadn't worsened. She was awake a little more often, but she seldom ate or drank anything, and she needed to be gathering strength for her upcoming labor.

As Avi slept, Nuri put her ear to her friend's distended abdomen and smiled thinly.

"At least the baby's heartbeat is nice and strong," she said to Lehoy.

Raising her head and frowning she whispered, "But it's been ten moons, so the baby will come soon. Really soon. I just hope Avi can handle that."

These days, Nuri spent most of her time tending to Avi, but she also helped Sima in caring for Nindai. And there was good news with him. He was responding nicely. The swelling in his leg was way down, and he had much more energy than a few days before. Even Jelli looked less exhausted, although right now she was dozing against the back wall.

"Avi, you need to get better," Nuri said, knowing the sleeping woman couldn't hear, but wishing it with all her aching heart. "I can't…can't lose my best friend. And Lehoy's here, and he's praying for the same thing, too."

Inside, Nuri was in turmoil, needing to be elsewhere, but constrained by her devotion to her friend.

Nuri saw Avi's green eyes flutter half open. She licked her lips with a dry tongue.

"Here, take a sip of this," Nuri said bending to touch a small, water-filled gourd to her lips. Avi took a sip, and Nuri smiled encouragingly. With her second sip, Avi coughed violently, then turned her head to the side.

"No more," she said weakly. "Don't think…" Avi's eyes closed. And she shook her head almost imperceptibly. "Don't think I can…" There was a long pause, and Avi continued, "I can't do this."

Nuri gripped her hand and said, "Yes, you can. We're going to get you through this." Avi gave her hand a weak squeeze, and then it went limp, and a breath escaped her.

"Avi," Nuri said loudly. "Avi!"

Putting her fingers to Avi's neck, Nuri felt for a pulse, her own heart pounding with worry. Finding no pulse, she cried, "Avi... Avi."

Lehoy knelt beside his mate and gave a low, mournful moan, his hands pulling at his hair.

But in that same moment, Nuri said, "Oh, Spirits—there it is. Her life force still beats. It's weak…it's tiny, but she's still with us. Thank you Spirits. Stay with us Avi. Jelli, wake up!"

The aged medicine woman's eyes opened briefly, then closed again. But she put her hands behind her head and stretched, producing soft popping noises from her back. "What?" she said, yawning.

"Jelli," Nuri said louder. "What do we do? Avi's pulse…it's weak, It's really weak. What can we do?"

Jelli was awake now, but she just grimaced and looked down at the ground shaking her head.

"Jelli, do something," Lehoy said more harshly. "Or tell us what we can do."

But Jelli bit her lower lip and shrugged. "I've tried everything I can think of except…" She was quiet for a moment, then blew out a breath. "Go get Nindai, I need to ask him about… Well, there's one other thing…" Jelli shook her head reluctantly. "Go get Nindai," she said again. "If he agrees, then maybe…"

Nuri was up immediately, and she ran to Nindai's hearth.

"It's Avi," she said in a loud, brusque voice, her head just past the hearth entrance. "She's so weak. Her pulse…can barely feel… Jelli says if you agree, we can try one other thing. Please Nindai, hurry."

Nindai was up in an instant, grabbing for his cane. "I think I know what Jelli's considering," he said. It's dangerous—kills about as often as it helps. But maybe…"

With Nindai hobbling as fast as he could, the two of them made it back to where Avi lay motionless with Jelli hovering over her chanting an appeal to the Spirits while Lehoy chewed on his thumbnail. In hushed tones, Jelli and Nindai spoke briefly, then both nodded their heads in agreement.

In a low voice, Nindai explained to Nuri and Lehoy. "There's a bluish-green mold that forms sometimes on our old meat or on fruit—like plums."

"I've seen it," Nuri said quickly, her eyes gleaming. "We cut that off and throw it away. Can that help Avi?"

"Well, sometimes it works in these kinds of things, and I mean it works really well. But other times…about half the time really…" Nindai 's whole face seemed to wrinkle. "Those times…well, they die."

Nuri's excitement drained away, and she pursed her lips looking to Lehoy.

"What do you think, Lehoy? It's got to be your call."

Nindai saw their misgivings and said, "Pretty much it's a last resort, but…if she's this weak, she's never going to make it through labor. I think Jelli's idea might work."

All eyes turned to Lehoy, and they saw him give an almost imperceptible nod. Nuri, too, nodded reluctantly. "That's what I was thinking," she said softly. "Seems like it's now or never."

Shortly thereafter, a recently recovered and energized Nindai helped Nuri search for some of the mold. There was none on the Nerean's meager dried meat stores, and they found no plums at all. A basket of crab apples had a few bad ones which they threw away, but there was no mold.

"Are those any good?" Nindai asked, pointing at the basket of small, hard apples.

"Well, they're really sour. They'll just about turn your face inside out when you eat them," Nuri said smiling tightly. Even as she smiled, she unconsciously pulled at her hair, worried that they wouldn't find any of the mold they needed.

It was in Moctu's mother, Alta's hearth where they found some. She had a hoard of bilberries she'd collected that had mostly spoiled, and they had a nice bloom of the mold.

"I found those a while back from some low-growing shrubs south of here," Alta explained. "They spoiled quickly, but I was saving them for their bluish-purple juice—great for dyeing skins. You think you can use that nasty stuff to help Avi?" she asked doubtfully. "Is that edible?"

"*All* things like this are edible," Nindai said. "Some are only edible once."

It was an old joke, but Nuri smiled nonetheless. It was usually followed by the admonition to "Take small bites, then wait."

That afternoon, with Jelli directing the procedure, Nindai helped Nuri prepare the concoction, and with excitement overridden by enormous apprehension, they woke Avi and got her to swallow about half of what they'd prepared. At nightfall, they gave her the rest.

Nuri chewed a thumbnail down to the quick during the night while she watched and attended Avi closely. Twice she prayed to the Spirits, one hand clutching the Earth Mother figurine and the other clasping her red hawk feather talisman. Lehoy also prayed, and he cut his hand to make a blood offering to the Spirits.

"Please Spirits, don't take Avi. *Please*," Nuri implored. "She means so much to all of us. We've already had so much misfortune. We're all so tired, and we need your…we need powerful action from you. Please help her get well, and…let her deliver a happy and healthy baby. Please let this new medicine work."

It was dawn as Nuri was awakened by commotion, having fallen asleep against a side wall.

"I'm hungry," Avi said loudly, trying to sit up.

Jelli was already up and preparing soup, her back to Nuri. When she turned, she gave Nuri a huge smile as Nuri's eyes came into focus.

"She's better," the little medicine woman said, her eyes glistening with tears. "She's much better."

"Oh, Spirits," Nuri said, placing a hand to her chest. Tears of relief streamed down her cheeks as she jostled Lehoy awake and gave him the good news. He was at Avi's side immediately, kissing her hand over and over. He looked to Jelli with glistening eyes, tried to speak, cleared his throat, then looked away to surreptitiously wipe an eye.

"Thank you Spirits," Nuri said, "And thank you Jelli. You're such a good healer. You saved her…You're—"

"I'm old," Jelli said simply.

"No, Jelli, you're… You just saved a life Jelli—my best friend. How did you know? How do you remember all that?"

The wiry, old woman smiled at the praise. "The afternoon knows what the morning never suspected," she said. "That's why older people can be a little wiser."

Chapter Sixty-Eight

"I'M SORRY," NARAMSIN SAID as he tied Moctu's hands behind him. We'll try to get Rolf to change Devu's mind, but he seemed pretty…set. The back chamber…it's very dark and the air isn't good."

"Why do you let him treat you like that?" Moctu demanded. "Why does Rolf let him lead?"

"He's very…powerful," Naramsin said carefully. He's destroyed men—strong men—with his thoughts and words. It's hard—"

"So, you just let him walk all over you…insult you in front of so many people?" Moctu said loudly.

Naramsin grimaced and quickly motioned both hands downward for Moctu to lower his voice.

"I'm supposed to gag you if you talk, so keep your voice low. There are some of us…many really, that worry about the fires. Your words made sense. But Devu—"

"He has such a hold on you all," Moctu interrupted. "How'd that happen?"

"He killed my uncle, Tebow—a very strong, healthy man—just by glaring at him. They'd argued earlier in the day, and at dinner Devu focused his eyes on him for a long time. Tebow laughed it off, but then he started to choke. He foamed at the mouth, then knelt down and died. It was years ago, but lots of people saw it."

"Could have been poison," Moctu said frowning.

Naramsin was quiet for a long moment then said, "No, I don't think so. It's not just that. All of…well, lots of his prophesies come true. He has real powers. He'll live forever."

"What?" Moctu said loudly, and Naramsin again hushed him.

"Even though he's old, he says he says he'll live forever. He's been told by the Spirits. He said they told him that he can't be killed by man or woman…or by rock, wood, antler or bone. He can't be killed—they've told him that several times. How can you fight that?"

"Give me a spear, and I'll show you," Moctu said scowling.

Naramsin gave a small snort and smiled tightly.

"You'd fail." Naramsin shook his head bitterly. "I'm leaving now, and you'll stay here. There'll be a young guard—probably two—at the front of the passage. You're not to leave."

"But what—"

"I've said too much already," Naramsin said quickly. "It's best that you stay here and not make trouble."

With that, he turned and left.

"Crap, now what?" Moctu muttered peering around the dark chamber. "That couldn't have gone much worse. Didn't get Elka released, and now I'm stuck here. Dark, smelly...like I'm rammed down a giant hyena's butt." He sat down to think.

"Well, at least we know now why there's so many burned-out areas. Not lightning storms, it's the crazy Lion warriors and Devu. And it was successful for a while—that's why they had 'more meat than they could use' like they said—they killed off the herds or drove away the few that were left." Moctu shook his head, and his eyes widened as another realization hit him.

"That's why they have so many people now—all the meat they could eat for years and years. And we didn't hunt that way, so we struggled while they had meat to waste."

He stood and found a rough projection on the cave wall, and he went to work sawing his binding against it.

Moctu scowled repeating Naramsin's last words. "Don't make trouble he said." The scowl slowly transformed into a thin smile, and Moctu nodded. "Oh, I'll make trouble alright."

Chapter Sixty-Nine

"You right," Pele whispered. "No good here for you. Pele can't fix. You stay, they kill. Need escape."

It wasn't among any of the statements Elka expected Pele might make, and her mouth opened, but nothing came out.

"Elka?"

"Uh, yeah," Elka answered uncertainly.

"We go—maybe this night."

Elka's head cocked as she tried to figure how he might be tricking her and why. She had to know more, but all she could manage was, "What?"

"We go—maybe this night."

Elka's confusion faded, and she scowled at the silhouette in the entryway. "Why should I trust…how can I trust a person who tried to kill my family?" She bristled as the next thought immediately came to her. "The man who killed my brother?"

"*What*?"

Even in silhouette, Elka could see Pele's head jerk back in bewilderment.

"What?" he repeated louder.

"You heard me—I know it was you." Elka was so angry that she spit toward him after her last word. "Don't deny it."

"No…no…that's—"

"And you almost killed me and Misha."

"No, no, I…I *save* you. You very sick and hurt. I save."

"I saw your red tunic—found it in your pack. Still had Misha's tooth marks on it," she snarled. "I staked it out to let you know I found it. Then I left. Should have killed you."

"Ohhhh," Pele said, as if a puzzle had been solved. After a pause, he gave a small snort then said. "I *trade* for tunic. I trade for…with Bullow. He's Lion warrior. Hunting now—not here."

Elka's eyes narrowed, and she frowned.

"But…" she said shaking her head. "Scars…scars and scratches—you had them on your left arm where Misha—"

"No…no. That from seal, wounded seal. We thought dead, but…no." After moment, Pele continued, "Elka, I…" Pele looked quickly to the right. "Oh, Adolfo—the guard comes," he whispered loudly, backing out of the entrance.

"Wait," Elka said, still baffled and needing more details.

"Thanks for watching her," Adolfo said, "I'll take over now."

Elka managed to drape the loose binding over her wrists just before Adolfo stuck his head inside the entrance to check that all was secure.

"Yes, I help guard more times," Pele offered. "No problem."

Elka heard him walk away to the right. Her head was spinning, so she sat down, trying to process what just happened.

"Spirits, could any of that be true?" She ran her hands through her hair and stared blankly at the ground. A host of questions came to her that she wished she had asked. "What about the other things in your pack?" she muttered. "The ochres, the minerals—the exact same colors the attacker had worn—red ochre and white chalk. And the moccasins with the hole on the ball of the right foot—exactly like the one I tracked of the killer. Did you trade for those too?" Elka shook her head. "I don't think so."

"But he did save us—both me and Misha. And Misha seemed to like him okay, even though she growled at the tunic. But she likes anybody who gives her caribou," Elka smiled at the thought of her wolf, but it quickly faded, and she bit her lip worrying about how Misha was coping.

Her brow wrinkled as another question came to her—a big one. "What about Mia?" she murmured. "That happened pretty quickly—too quickly. If Pele's just a simple trader, how did he get her? Why would they give such a beauty to a trader, and so quickly? And so many young warriors around here needing mates."

Elka put a hand to her forehead and looked down. "What did Naramsin say? That Devu gave him Mia because he wanted to keep Pele here?" Her mouth fell open as she realized that Devu just wanted Pele for his knowledge of how to make bows. And it had worked. Every Lion warrior she'd seen had a bow.

Her thoughts were interrupted by Adolfo who stepped inside the hearth and set a skin of water down. He sniffed the air and wrinkled his nose. "Stinks in here." Then his frown turned to a wide, smug smile. "Devu's busy right now—planning some...*sacrifices*." He paused to let the words sink in and have full effect. "Looks like it'll be a man, a girl...and a wolf." His smile broadened maliciously. "But he says he'll see you later this afternoon. I'll go check in a while to see if he's ready. Here's some water to drink...and to clean up with." Sniffing one more time, he shook his head, snorted and left.

"Spirits," Elka muttered, covering her face with both hands. Although she already recognized the likelihood of the three sacrifices, the actual words took her breath away and sent a tremor through her. Devu's suggestion of canceling or delaying the sacrifice echoed in her mind, and as she was pondering what she should do, another memory came to her.

It was five or six years earlier when two boys about her age, Itzal and Indar, were playing with a long-tailed rat down by the lake. Both of the boys' fathers, Jabil and Jondu, were dead, and the two boys—especially Itzal—got into more than their share of trouble.

"Watch this," Itzal said to Indar as Elka watched from some distance away. "If I hold his tail, he turns to try and bite me." Itzal smiled as he expertly jerked the rat's tail every time it got close to biting him. "But now," he pulled the tail up so the rat was dangling, and he walked to a deep pool on the rocky edge of the lake. "Now, if I hold him in the water..." Still holding the tail, Itzal submerged the rat in the water for a long moment, then allowed him to come up for air. The rat coughed, squealed and frantically paddled to keep from going underwater again. "He doesn't even *think* about biting me." Itzal paused as the desperate animal's movements became more frenzied. "You know, if he dove down to bite me," Itzal continued, "I'd probably let go, and he'd get away. But he's only thinking about his next breath."

The boys had continued tormenting the rat until it was dead, despite Elka trying to intervene. Later, when she told her mother about it, Nuri had frowned and said, "Boys…some boys are like that." She paused and shook her head. "And it's those kinds of things that…that help you decide who to stay away from."

But Elka nodded as the realization set in. "Right now, I'm the drowning rat," she murmured.

Chapter Seventy

NURI GASPED FOR AIR on an uphill stretch, pushing her body to make the journey in a day less than usual. She was sweating profusely despite the near-freezing temperatures.

"Moctu's in trouble—I can just…I just *feel* it," she murmured, trying to give her body incentive to fight through the exhaustion. "And it's less than a quarter moon until it's completely full," she worried, thinking of Elka. "Got to push harder."

She allowed herself a faint smile, thinking of her sister-friend, Avi. "At least we got you through your suffering, and we helped you deliver a healthy baby girl. You really had us worried there."

Her smile broadened as she pictured baby girl Zirtea, which meant 'lucky' in the old language. "That name fits you so well. Your mother went through a lot having you, but in the end, you both came through just fine. Baby girl, I hope you're lucky your whole life. But let's don't call it so close next time."

As she pressed herself to move faster, Nuri had another thought. "Zirtea—maybe you can give us a little of your luck. We need it. Going to be tough to find Moctu in all that wilderness around the Lion People camp. They might even have him prisoner by now." She shuddered, imagining some of the awful things they might do to Moctu if they caught him.

"And Moctu was going to try to speak to them under a branch of truce. Let's hope they honor it. But knowing Devu…" Nuri frowned, not finishing the thought.

Unconsciously, Nuri was now jogging along the trail.

The uphill path gave way to a broad, meadow of tall grass and small shrubs, and Nuri jumped back as she startled a large flock of doves who arose in a thunder of beating wings. Her heart had already been beating hard, but with a surge of adrenaline, it felt as if it were coming out of her chest. In spite of the urgency she felt, Nuri came to a full stop with a hand to her heart.

"Got to go a little slower. The birds may have attracted big cats like leopards."

But the ground was now flat, and soon her worries about Moctu and Elka had her jogging again.

Chapter Seventy-One

"LET ME TALK TO her," a deep voice said to the guard.

"Okay Bullow, but don't go in. I have orders—"

"By the Spirits, I'll go in if I want to," Bullow said belligerently, and he thrust his head inside the hearth entrance.

At the mention of his name, Elka came out of a daze of confused thoughts, and she stood, wanting to see if Bullow could be the man who attacked her family and her as Pele claimed.

The silhouette in the hearth entrance was indeed tall with broad shoulders, as was Pele, and Elka's eyes narrowed as she tried to see more of Bullow's features.

Bullow gave a short laugh. "They told me you were here, but I had to see it with my own eyes. I should come in there and twist your head off," he said. "You killed Tommu, my best friend. And your filthy hyena almost tore my arm off."

Elka smiled as her confusion and worries receded, and a deep glow of satisfaction spread through her. At least for this brief moment, she was happy—suffused with triumph and validation. Her initial feelings about Pele—and his for her—had been genuine. The love she had felt, and his affection for her had been real. She hadn't misread him. And moreover, her quest to avenge the attack on her tribe and family had been successful, at least in part.

"Thank you Spirits," Elka murmured. She cast her eyes upward briefly. "That was for you, Meko."

"Yeah, you'd better pray you little bitch," Bullow said, misconstruing her actions. "I searched for you, but I never could find your trail," he mumbled bitterly almost to himself. "I knew I wounded you, and I thought I'd killed your stinking hyena."

His voice brightened from its sour tone, and he stepped in farther. Elka could see by the light on his shoulder that he was wearing a red tunic. It was different from before, but it was red.

"But I see she's missing an eye," Bullow continued. "And she's penned up waiting to be sacrificed. She's already miserable, and I'm going to make sure that no one feeds her 'til…" Even in the dim light, Elka could see the gleam of Bullow's teeth as he sneered, and his lip curled upward. "…'til we burn her. The full moon's only two days from now. And after the wolf, it'll be your turn…then your father's."

The warmth inside Elka dissipated quickly, replaced by worries about her father and Misha. Bullow chuckled, seeing the impact of his words.

"And I'll be there watching you burn and hearing you scream. We'll try to make it slow. I just love the smell of roasted Pale Ones."

The last comment greatly amused him, and Bullow laughed loudly. Still chuckling, he turned and walked away. Chilled by his comments, Elka wrapped her arms around her belly.

"Oh, my baby," she said softly. "Your father…well, he wasn't… *isn't* what I thought he was. And you're dearer to me than ever before. *Got* to protect you. And there's Father and Misha…Spirits, please help me. How can I save…" Elka didn't finish the sentence, and she stared at the cave floor, her mind swarming with worries and doubts. Every possible solution had insurmountable obstacles.

Then she smiled and cocked her head, thinking about Pele's comments. "Need escape. We go—maybe this night."

She'd dismissed it at the time, but now, it actually seemed possible.

"Could that be true?" Elka muttered, her eyes narrowing. "It has to be. But what'll happen with Father and Misha? That's never going to work. But my only other option is Devu, and what he wants." She shuddered and fought off the urge to spit. "And that's not going to solve anything. It just gives us more time like this—sitting in a cave waiting to be…burned."

Elka shook her head, trying to rid herself of the images that comment engendered. But another thought came to her, and she nodded, temporarily sidetracked.

"A year ago, this…what's happening now…it would have brought on the sadness times," she murmured. "But I don't feel any of that." She cocked her head in wonderment. "I just feel like…" The image of Bullow laughing came to mind, followed by one of Devu leering at her as he offered an extra moon or two of life in exchange for sex. Her mind recalled Itzal holding the rat under water and saying that it was only thinking of its next breath, not of biting. Elka's lip curled, and she nodded as she finished the sentence. "*Killing someone.*"

Chapter Seventy-Two

"I NEED SEE HIM," said a deep voice to Moctu's guard.

Moctu cocked his head, as the voice was strange—foreign, and not someone he recognized.

"He kill my brother, and I..." Moctu heard a fist smack an open palm. "You know."

"Okay, but just body shots...nothing on the face. Not supposed to...just don't hit his face."

"Yes, no face, no face."

Moctu stood, quizzically looking at the cave entrance, wondering who the foreign person might be. He had earlier severed his wrist bindings by rubbing them against a rough projection on the cave wall.

A large man with broad shoulders rounded a bend in the cave and strode toward him.

"I've never seen you before in my life," Moctu said as the man neared.

The powerful-looking man said "Make noise," and he slapped his hands together. When Moctu just looked bewildered, the man grimaced, his teeth clenched, and he hissed more insistently, "Make noise." The he smacked his fist into his hand again.

This time, Moctu nodded deciding to go along with the ruse. He didn't know who this man was, but he was deceiving Moctu's enemies, so he'd play along for a while.

"Oww! What in the Spirits...?"

The big man smiled and nodded, and hearing the guard coming, he hit Moctu in the stomach.

Moctu could tell that the stranger held back on the punch, but it still hurt, especially to his ribs. Moctu doubled over, groaned and coughed.

"Don't hurt him too bad, Pele. And nothing to the face," the guard cautioned.

Moctu immediately froze in confusion, his mouth falling open.

"So, this is Pele? He's the one..." Rage swelled inside Moctu, but it was slowed by the newcomer's strange behavior. Still doubled over, Moctu's lip curled in a snarl, and his body tensed, readying for an assault on the tall man. Neither the guard, nor Pele knew he was no longer bound, so his attack would be a surprise.

"No kill, just hurt. This for my brother." Pele called to the guard.

This guy attacked my family and killed my son, Moctu thought as he raised up to begin his charge.

Pele hit Moctu in the middle of his chest with more force than the last time. Moctu didn't have to fake pain, as the impact stunned him and took his breath away.

"Okay, but don't kill him. I'll get in big trouble." the guard said, turning to walk back to his position around the corner.

Pele grabbed Moctu by the shoulders, and in his ear, he hissed, "You make noise, I talk."

Still sucking for air, Moctu frowned and tried to shake himself free.

"Make noise," Pele said more insistently. "I talk."

Another ruse, Moctu thought perplexed.

"What the..."

Pele shook him again and said, "Noise."

Moctu nodded and moaned.

"I Pele," the big man said, as he hit the side of his own leg, making a satisfying 'whump' sound.

Moctu moaned again, his rage now outweighed by curiosity.

"Want save Elka, but..." Pele searched for the right words. "But I know...um, she not leave you here."

Moctu struggled to follow Pele's words, as the big man's grasp of the language was rudimentary, and most of his pronunciation followed the Lion People dialect. But Moctu understood and nodded, very intrigued.

"Make plan, but…hard." Pele frowned and distractedly rubbed the back of his head. He then realized that he hadn't pretend-hit Moctu for some time, and he hit the side of his leg again.

Moctu groaned, then whispered a rush of questions. "What are you here for? Why would you help us? Why'd you try to kill my family? Why…"

When he saw confusion on Pele's face, he tried simplifying his speech.

"You killed my son, Why I trust you?"

Immediately, Pele frowned, shook his head fiercely, and said, "No, no, no. That not me. That Bullow." He shook his head forcefully again, "That not me. I no kill. Get Bullow red…" He couldn't think of the word, so he grabbed his own tunic and pulled on it. Get in trade."

Maybe it was because Moctu's options were so limited, that he *wanted* to believe this stranger. But Moctu frowned, realizing with surprise that he did.

"Spirits, I may regret this, but…" Moctu touched his chest, then pointed to Pele. "I think I believe you."

Moctu shook his head, then decided to tell Pele more. Before saying anything, he hit his own leg and groaned.

Pele smiled and nodded.

"Have you talked to Elka?" Moctu asked.

Pele nodded and held up a hand with his thumb a small distance from his index finger.

"Little talk," he answered.

Moctu nodded and whispered, "Did she tell you…" Moctu started, then simplified it again. "Elka carries your child." To make sure Pele understood, Moctu said, "Elka…" He then patted his belly, and thereafter pantomimed a much larger belly. "Your baby."

Moctu saw surprise in Pele's face, and he knew the man was hearing new information, and that he understood. Pele smiled, looking thoughtful for a moment, then he smiled more broadly.

Moctu decided to reveal his only other possibly helpful piece of information.

"Naramsin might help. He seems…" Moctu saw confusion on Pele's face and realized he was having a hard time following the words. He simplified by saying, "Naramsin no like Devu. Maybe help."

Pele pursed his lips and nodded slowly.

"Need thinking. We go not tonight. Maybe tomorrow night."

Moctu nodded, and Pele smacked his fist into his hand, then turned to leave.

"Uff," Moctu said loudly, then moaned.

"I said no hitting the face," the guard commanded loudly, reentering the deeper part of the cave.

As Moctu sank to his knees, Pele replied to the guard, "I no kill. I go now."

Chapter Seventy-Three

The next morning the frigid wind blew hard, gusting through gaps in the heavy skins that shielded the hearth entrance. Numb with cold, Elka huddled in a back corner of the hearth, stretching her tunic to cover a portion of her face. Adolfo, the guard was shivering as he came inside the hearth and tried to close off some of the larger openings. He turned, wrinkled his nose and scowled at Elka.

"Stinks in here, but it's too cold outside."

"Yeah, the smell's much worse since you came in," Elka said.

Adolfo glowered at her then smirked. "Little hyena bitch. Get your smart comments out now, because the moon's full and all you'll be doing soon is screaming. We'll be turning Sleeps With Wolves into Burns with Wolves." He snorted then laughed at his joke.

Elka's brow knitted as her brave veneer crumbled for a moment, but she recovered quickly.

Need to talk with Pele, she thought. Need to get more time. Father, Misha and I are all in separate places, so we need a plan. Going to be hard to make this work. Impossible maybe. Elka shook her head trying to rid her negative thoughts.

"Adolfo?" Elka heard Pele say from outside the hearth. "You cold? I bring extra furs. You want?"

"Oh, yeah," Adolfo said. "That would be great."

Elka's pulse quickened, and she knew this was probably her best chance to talk with Pele. But she was disappointed when Adolfo only opened a small, lower corner of the entrance skins to accept the furs leaving Pele outside.

"Thanks Pele, I owe you. I'm freezing my butt off."

Adolfo began to pull the skin closed again when Elka blurted loudly, "Tell Devu I'm ready. Go tell him."

With a double take, Adolfo looked at her, his mouth open. "You want to be sacrificed now?"

"No," Elka said. "Tell him I don't want to die at this moon, and I'll meet him tonight. I'm ready for what he has in mind. He'll know what I mean. Tell him."

Adolfo studied her for a moment, then pursed his lips.

"Alright. If Pele will guard you for a little while, I'll go tell Devu, but this better be something he wants."

"Oh, he wants it alright," Elka replied.

"Pele—you good to guard her for a little while?"

"Yes, I guard," Pele said, maybe a little too quickly.

Elka eyed Adolfo, but he hadn't seemed to notice, and he readied himself to go out in the cold wind as Pele entered the hearth.

"Tell Devu that I'll come to his…that you'll bring me to his hearth when the Seven Sisters are at their highest point in the sky. That'll make for the most favorable meeting time."

Elka caught Pele's eyes to ensure that he heard and noted the time. Pele nodded slightly, indicating that he understood.

"Crap, it's cold. I hate going out there," Adolfo said wrapping the furs tighter around his shoulders. He scowled at Elka and said, "This better be important."

"I'm sure Devu will think it's important," Elka assured him.

Adolfo took a breath then scrambled past the hearth skins and bolted to the left toward Devu's hearth.

Pele and Elka rushed to each other, and Elka hugged him tightly.

"I believe you. You didn't hurt my family. I'm sorry…sorry I doubted you."

Pele hugged her back and said, "No much time. Adolfo back soon. We go tonight—yes?"

"One more thing," Elka said. She patted her stomach. "I carry your baby. You should kno—"

"Moctu tell me," Pele said smiling and nodding. "Ver happy."

"What about Mia?" Elka asked.

Pele pursed his lip then cleared his throat. He opened his mouth, but couldn't think of what to say.

"Never mind, we'll deal with all that later," Elka said hastily. She smiled, glad that Pele was pleased, but she had a lot of questions about Mia. "We need a plan now—and quick. So, Devu wants to… he wants sex—tonight—and I don't plan to…" Elka saw confusion on Pele's face, so she simplified her words. "I not have sex with Devu this night. He be mad. Going to fight. After fight, we go then. With Father—Moctu—and Misha." She pointed to Pele. "You need free Misha from cage, and get Moctu. You understand?"

Pele nodded, but he looked troubled. "Ver hard," he said.

"Yeah, but this is our best chance, probably our only chance."

"Okay, I free Misha. You whistle—maybe she come?"

"Spirits, is it cold out here," they heard Adolfo exclaim from outside as he returned.

Hastily breaking apart, Elka again huddled in the corner as Pele helped open the hearth entrance skins for Adolfo.

"Any trouble?" Adolfo asked.

"No trouble," Pele said shaking his head.

"What did Devu say?" Elka asked.

Adolfo paused and scrutinized her. "I'm to take you to him when the Seven Sisters are at their highest. Like you said," the tall guard grumbled, "He seemed to think it was important."

Chapter Seventy-Four

IT WAS A CLEAR night, and the full moon shimmered in a dazzling sky covered by countless, twinkling stars. Although numb from the bone-chilling wind, Pele easily located the three-starred belt of the Great Warrior. From there, a short distance to the right he found Taurus the Bull who carried the Seven Sisters on his shoulder.

After studying them for a long moment, he muttered in his language, "Don't have much time." Looking down, he rubbed his forehead and tried to contrive a plan.

"That'd get us all killed," he muttered discarding his first idea. Pele's hand slipped to rubbing his short beard, but still, no viable plan came to him. Shivering hard, he said, "Got to get warm. Can't think." Momentarily suspending his effort, Pele hurried to the hearth he shared with Mia so that he could stand near their fire and think.

"You're ice-cold," Mia exclaimed hugging him tightly. Pele hugged her back, savoring both her warmth and the captivating curves of her body. The doe-eyed beauty pulled him closer to the hearth fire and said, "Where have you been? I've been worried. It's so cold out there."

And what do I do about Mia, Pele wondered bleakly. Devu gave her to me in payment for me teaching them about bows. And she's beautiful and…really wonderful, but…He shook his head. Too many things to think about. Have to save Elka and her father—can't let them die. But what's Mia going to think about all that? I think she'll come with us, but I don't know.

"You're so quiet. What's troubling you?" Mia said disengaging from him.

"Just need thinking," Pele said staring into the fire.

"Maybe I can help you think," Mia offered.

Pele continued to stare into the fire, but shook his head slightly. "No...thanks."

Irritated and a little hurt by Pele's unresponsiveness, Mia withdrew from the fire to work on a late dinner. "You hungry?" she asked. "I need to know how much dried caribou to put in the stew."

The mention of caribou did it. Pele had a plan. Not a great plan, but one that wasn't doomed from the start.

"Yes, make some for me please, and little more too."

"Oh? Who's getting the extra?" Mia asked.

"I take to Moctu and his guard," Pele replied carefully. "He...they both ver hungry and cold."

"You think that's a good idea? Feeding the prisoner?"

Pele nodded, but didn't meet her eyes. "Both ver hungry," he said again.

Mia's eyes narrowed, but she kept her tone upbeat. "I'll make a nice, extra skin of stew for them."

She frowned as she went back to work. "Something's up," she muttered quietly.

Chapter Seventy-Five

"Spirits, I'm cold," Moctu said pacing with his arms crossed in front of him. Even in the dim light he could see his breath fog. He had asked for furs, but was told that Devu forbade it.

"Haven't been this cold since we hunted the kroag-kroag up north." Despite his current misery, Moctu smiled at the memory of hunting mammoths with the Pale Ones, including his friend, Hawk. He had never been that cold, shivering until the beasts came, then being nearly incapacitated with fear by their horrifying enormity. But they'd been successful and killed two mammoths, a huge amount of life-saving meat to help the clan make it through a harsh winter. Ever since that time, Hawk had called him Brud Moctu, which meant Brother Moctu. He, likewise, thought of Hawk more as a brother than a friend.

"One of the finest men I've…ever known," he murmured. Moctu shook his head sadly, wondering if he'd ever see his Pale One brother again. He'd considered enlisting Hawk and the Krog to help in their battles first with the Sea People and now with the Lion warriors. But the Pale Ones had no bows or even atlatls and they'd be no match for the well-armed Lion warriors. They'd just be slaughtered.

Moctu's teeth began to chatter as his ruminations shifted to Nuri, and he wondered the same about her. His breath caught at the thought of never seeing and holding her again, and he felt an immense hollowness in his chest.

But gradually, his demoralized emptiness shifted into anger.

"I came here during a truce, *guaranteed* by those four warriors. Could have killed them, but instead, I trusted their promises." His lip curled, baring his clenched teeth.

"On their honor," he muttered, then spit. "The Spirits curse them and their *honor*," he said louder, spitting again.

Moctu had been kept in the cave for a long time—a couple of days anyway, all without food and only a little water. He regularly edged around the curved cave passage to check out the guards. There were almost always two—young, strong and wielding spears with razor-sharp flint points. He'd held off from attempting escape, not only because the odds of success were low, but because his escape would mean certain death for Elka.

But now, he was freezing, and he was hungry. The moon had to be full by now, so they'd sacrifice her, and probably him, too, very soon.

"Not much to lose," Moctu muttered, gearing up for his attempt at escape.

"I see Moctu now," came a foreign voice that Moctu recognized as Pele's. "Give stew caribou."

"Devu says no food," said the larger of the two guards. He was young, but very tall, with two long braids that were tied with white strips of leather making him look vaguely feminine. His voice was deep, however, and he waved his spear with authority.

"Devu say this alright," Pele lied. "Since sacrifice coming."

"Well maybe, if Devu—"

"I go in now." Pele interrupted, brushing past the guards.

He rounded the curved passageway, and Moctu, who could already smell the stew, met him just out of sight of the guards.

"So hungry," Moctu said, eagerly taking the skin of stew that Pele extended toward him.

"We go this night," Pele whispered as Moctu held the skin to his mouth and began half-eating, half-drinking the thick mixture of caribou, onions and mushrooms. But even as he devoured the stew, he nodded, confirming that he understood.

"We go when Seven Sisters are highest in sky. You can't see, but listen. Will need this," Pele said, giving Moctu a long-bladed flint knife well-hafted to an oak handle.

Having just finished the delicious caribou stew, Moctu smiled at the sizeable weapon and hefted it approvingly.

"I open cage of wolf Misha. Dark and cold, maybe no one see. But maybe someone yell. Then, I come help you with guards here. Elka with Devu and his guard outside—Naramsin. We go there quick quick. Help Elka."

Taking off his fur, he wrapped it around Moctu's shoulders. Unexpectedly warm, and with a belly full of stew, Moctu felt a rapidly growing fondness for this foreigner whom, just a few days before, he had hated and wanted to kill.

"What's taking so long?" the big, braided guard said crossly, entering the deeper part of the cave, his spear at the ready.

Pele stepped in front of Moctu, blocking him from the guard's sight as Moctu quickly hid the knife in his tunic.

"He finish stew. I take skin now. Leave now," Pele said, moving directly toward the upraised spear as if he felt no worry.

Satisfied, the guard pulled his spear aside, and escorted Pele outside the cave. Pele stole a look upward at the Seven Sisters, then said, "You guards hungry? Maybe I bring stew soon for you?"

"I'd like that," the big guard said, looking at the other guard and getting confirmation. "Yeah, we'd like that—the hotter the better. Thanks."

"Back soon," Pele said walking away.

Chapter Seventy-Six

"HERE'S THE LITTLE BITCH," Adolfo said to Naramsin who, bundled in furs, was trying to stay warm outside Devu's hearth. Looking up, Elka's guard continued, "Seven Sisters are as high as they're going to get."

"Okay, Adolfo," Naramsin said. A spear in one hand, with his other one, he drew Elka away from Adolfo and nearer to him. "Thanks, I'll take over from here."

When Adolfo stayed where he was, the scar-faced older man followed up in a more commanding voice, "Thanks Adolfo, I have this. You're dismissed."

Reluctantly, Adolfo turned and slowly walked away.

Naramsin waited until Adolfo was a handful of paces away, then he looked at Elka and whispered, "I'm sorry that…sorry you have to do this."

"Yeah, me too," Elka said glumly, but inwardly, she resolved that Devu wasn't going to be getting what he wanted this night.

'I wish—" Naramsin started.

Glancing upward to the stars, Elka quickly cut him off. "Let's just get this over with."

"Naramsin, is she here?" called Devu from inside the hearth.

"Yes, Devu, she's here," the old warrior said while opening the hearth entrance skins.

"So, here you are," Devu said, smiling at Elka and rubbing his hands together. "You've decided to be purified. That's good."

Elka noted that the boney, old man's beard was clean and combed, and his scraggly hair had been recently washed. He was wearing a robe, tied at the waist with a narrow leather belt. There was a modest fire burning, but the interior was comfortably warm.

In a large seashell on a flat boulder, some dried vegetable matter was smoking, giving off a fragrant aroma.

"First, we both need to drink these," he said, holding up two more shells that each had puddles of dark brown liquid. "Don't worry—it's nothing bad… You choose one, and I'll drink the other. That way, you know that neither is poison."

Elka looked at the brown liquid doubtfully, but she chose the one nearest her.

"You drink first," she said.

Devu slurped the contents of his shell bowl down, wiped his mouth and said, "See? Not poison."

Still frowning at the mixture in her bowl, Elka said, "So, it sounds like you know a lot about poison and what's not poison."

Outside, Naramsin huddled close to the hearth skins which radiated enough heat to slightly warm him. The wind was gusting hard, and his fur hood was drawn tightly to his head, partially covering his ears and face. His ears ached with cold, and Naramsin occasionally pushed them inward with his gloved hands to warm them and relieve the pain. He could only hear a few words coming from inside the hearth, but at the mention of poison, his head cocked, and he drew even closer to the entrance skins. Pulling the hood away from his ear, he positioned it as close as he could to try to hear the conversation inside.

Devu gave Elka a knowing grin, then said, "I've used poison sometimes in the past." He touched his lower lip that was growing numb from the concoction he'd just swallowed. The mixture was a puzzle, teasing him into wanting to talk more, but making speech more difficult as the mouth got numb.

Gesturing to Elka's bowl, he said, "Drink up, drink up."

Elka knew the mixture was Jin or something like it, and although she had never had it, she knew of Jin's effects. Jin was a mushroom that sometimes sprouted near caribou droppings. When consumed, it allowed people to peer into the Spirit world, but it also made them talkative and thereafter, sleepy. She'd never had it because the Nereans restricted it to men who had been through their manhood rites and women who were mated.

If it makes him sleepy, that could help, Elka thought. I'll play along and have a little. I've always been a little curious.

She took a small sip of the mixture, then Devu put a finger under the bowl pushing it upwards, causing her to swallow about half of the contents. Almost immediately, her mouth went numb, and Elka realized she was supposed to be whistling for Misha about now. That was going to be hard. She tried forming her mouth into the shape needed for whistling, and she managed a softer-than-usual "Suureet–suurr," drooling a little at the end.

When Devu looked surprised, Elka wiped her mouth and explained, "Mouth is all numb. Wanted to see if I could still whistle. Suureet–suurr," she whistled again, somewhat louder.

Devu smiled and moved closer and looked into her eyes.

"Your blue eyes are very different…and beautiful."

From any other man in almost any other setting, Elka would have appreciated the words and considered them sweet and charming. But from this scrawny skunk of a man—a man who had the power to send her to her death and would likely do it soon—the words seemed hollow, even treacherous. Elka knew that was the case, but her mind was clouding, and she was having trouble connecting her jumble of thoughts. She had no idea of what to do now, and she stood there dumbly.

When she made no reply, Devu continued in a husky voice, "Now it's time for purification." He touched her shoulder, ran his hand softly down her front, then began unfastening her belt. His robe fell open revealing a rigid penis that looked oversized for his spindly body. Elka stared at his member for a moment before he noticed her gaze and grabbed her hand.

He directed it downward, saying, "Now take this with your hand…"

That's when the pandemonium started.

Chapter Seventy-Seven

It was completely dark outside, but the fire's embers were still bright, and the hearth was comfortably warm. Mia feigned sleep, but her left eye opened a slit as Pele silently arose from their sleeping furs. After donning his furs, Pele collected the stew bag with its cold remnants, and he quickly and quietly left the hearth.

Her suspicions aroused, Mia got up, threw on her boots and furs, and she stealthily hurried after him.

Her lip curled as she murmured, "He better not be taking that food to his dear, little Sleeps With Wolves."

Somewhat relieved, Mia saw him head to the right, not the left, so he was going somewhere else. After he walked past several more hearths, she muttered, "Nothing down this way but that stinking wolf that whines so much."

As if on cue, Misha let out a long, mournful howl.

"What the…?" Mia muttered as she saw Pele stop at the sturdy cage. He fished in the stew bag for a large lump of caribou, and he pushed it through a gap in the wooden posts. There was a flash of teeth and the morsel disappeared quickly.

Mia almost called out to Pele, but she was curious now, so she held her tongue.

There was a far-off sound of a whistle, almost drowned out by the wind, but Misha's ears perked up, and her head swiveled in that direction.

"Okay Misha," Mia heard Pele whisper, "You free."

"Elka," Pele said louder. "Misha, you go Elka now."

Mia's jaw dropped, and she stared in disbelief as Pele unlatched the door to the cage, and the quickly stepped back as the huge beast broke from the cage.

"No!" Mia cried.

Pele turned to see her, and for a moment their eyes met.

Pele put his hands out, urging her to silence, saying "Mia, please. Mia—"

But Mia frowned at him, her lips pressed tight. She shook her head slowly, and her eyes hardened.

"He's letting the wolf go!" she screamed. "The wolf is out."

Pele's mouth fell open, and he moaned, "Oh, Mia—no, no."

"Pele…you, you…betrayer!" Mia said, almost spitting the words. She hit him hard on the chest and screamed, "What are you thinking?" As there was stirring in the hearths behind her, Mia continued to shriek. "Shoot it. Stop the beast. Don't let it get away."

Pele grabbed her and put his hand over her mouth.

"Mia, stop—please. Shh-shh-shh."

But it was too late. A young warrior wielding a spear appeared, and seeing the fleeting image of the escaping wolf, shouted, "Pele, you filthy snake." He raised his arm to throw but paused, not wanting to hit Mia.

Mia bit down hard on Pele's hand covering her mouth, sinking her teeth to the bone of his middle finger.

"Agggg!" Pele cried, releasing and spinning her to the side. Seeing the warrior begin his throw, he dodged to the right. But Mia wouldn't let him get away, and she held to his tunic. As she was dragged to the right by Pele, the spear took her in the cheek, breaking teeth and passing through the back of her head. She was dead in an instant.

Pele's eyes widened in horror, and he let out a primal scream.

"Agggg- nooooo!"

He bent to hold Mia but knew immediately that she was dead. Seeing the warrior stringing his bow, he charged the young man, roaring with fury.

"Grrrr-aaaag!"

Just as the warrior nocked an arrow, Pele's head and shoulders hit him on his right side, bashing him to the ground. His teeth bared, Pele vented his wrath and anguish, hammering the man's head with powerful blows. The warrior was unconscious, and Pele would have beaten him to death if two more warriors hadn't arrived.

"He kill Mia…and wolf is free," Pele said to the men, picking up the fallen warrior's bow and arrow quiver. "I go."

The confused warriors rushed to check Mia as Pele bolted in the other direction. As he ran, Pele moaned and squeezed his forehead.

"Ooooh Spirits…No, no, no. Spirits, why…why?" He had only known Mia for a few moons, but she was a good mate, and he loved her.

"I should have…Oh Spirits, I should've…" Pele shook his head trying to think how he could have done things differently. But he could think of nothing. His thoughts kept coming back to Devu. In his native language, Pele muttered, "Devu. He caused all this." His eyes narrowed, and he scowled. "Devu," he said once more.

As he approached the cave where Moctu was confined, Pele saw two tough-looking guards, each sporting a spear, and his attitude shifted.

"Concentrate Pele, this isn't going to be easy," he whispered.

At the far end of the shelter, he heard a tremendous commotion, and it drew the attention of the guards.

"That's probably Misha," Pele muttered, allowing himself a thin smile as he imagined the look on Devu's face as an angry, savage beast charged into his safe, warm hearth.

Chapter Seventy-Eight

Elka heard the whump of impact from just outside the hearth.

"Ooaagh!" Naramsin cried as he was bowled over by a giant beast hitting him in the chest. He was slammed to the ground, and his spear clattered down the slope as Naramsin used both hands to fend off a snarling, biting she-wolf from his face and neck. Even though Misha savagely tore at his arms and shoulders, Naramsin's heavy coat protected him from serious injury.

Elka's mind was no longer connecting thoughts. Rather, she had a confused maze of swirling impressions, of danger, longing, hunger and…love.

"Misha?" she said cocking her head.

She heard Devu yell, and she realized she was holding his manhood, and he wanted to…what…?

She frowned confused that Devu wanted to hurt someone or something.

"Misha," she said again, realizing Devu intended to hurt her precious wolf. Anger filled her, and Elka's nostrils flared. Her grip tightened on Devu's penis, and she pulled hard toward her at the same time headbutting his face. Two of Devu's teeth broke as they cut into her hairline, and she heard the satisfying crunch of his nose breaking. He was falling backward as Elka released his member, and his fall was accelerated as Misha smashed into his chest. Devu crashed to the stone floor of the hearth and lay there unconscious as Misha mauled his arms and face.

"Misha," Elka said happily, hugging the wolf from behind. Sensing that Devu was no longer a threat, Misha relinquished her hold on Devu's limp forearm, and turned to greet Elka. Her whole hindquarters shimmied back and forth with happiness as her dark tail wagged furiously. The huge beast whimpered with delight at the reunion, letting out little yelps as she quivered and bounced with excitement. She gradually calmed and began licking Elka's bloody forehead until Elka waved her off and stumbled out the hearth entrance.

Naramsin had been unable to find his spear in the dark, and he drew back as Elka and the wolf came out of the hearth. Elka heard him say something, but it made no sense as she and Misha trudged to the right toward more commotion.

Despite Pele thinking that overcoming the two warriors guarding Moctu would be difficult, it was as easy as drawing his bow and ordering them to put down their spears. Moctu was out immediately, brandishing his knife, and ready for action. Surprised, and a little disappointed at the lack of a struggle, he found an unstrung bow leaning next to the largest guard, and he quickly cut the bowstring and tied their hands and feet.

Behind them, the warriors examining Mia's corpse began shouting, and there were stirrings in several more hearths.

"Gotta go," Moctu said. "I hope Elka's okay. Knowing Devu…" He shook his head and scowled. He grabbed the furs that his guards had used to stay warm, and he and Pele began hurrying toward Devu's hearth at the other end of the shelter.

Warriors began emerging from hearths, wielding either dangerous-looking spears or bows, but looking sleepy and bewildered, not knowing the cause of the turmoil.

About midway to Devu's hearth, Pele saw Misha, and tottering beside her was Elka coming their way.

"It's Elka, but…she wounded?" Pele whispered to Moctu.

"Yeah, something's wrong," Moctu said worriedly.

As they met, Elka stumbled into Moctu's arms and said, "It's all turning."

"Are you alri—" Moctu began, but a nearby warrior shouted, "The wolf is loose. There…over there, with those…Who are those…?"

"Time to run," Moctu whispered loudly, and followed by Pele, he guided Elka into a shambling run down the slope away from the shelter.

It was fortunate that, knowing they'd need nearby firewood, the Lion People had not burned the forest bordering their shelter during one of their hunts. By the time their warriors realized Elka had escaped, the small, fleeing group was inside a bushy stand of young trees.

"She's escaped!"

"Sleeps With Wolves is gone!"

"Devu's been hurt."

"There were some people with the wolf—going that way!"

"Who's with her?"

As Moctu, Pele and Elka stumbled along a dark path in the forest, Moctu said. "Doesn't sound like they're following us yet… that's lucky. The nighttime and cold are helping us…taking them time to wake up and organize a group to come after us."

As a limping Misha stayed close alongside her, Elka tripped on a root and almost fell.

"You wounded…you alright Elka?" Pele asked grabbing her under an arm.

"I'm…" Elka said, in a puzzled voice, then was quiet for a long moment. "Thinking."

"You're acting really strange—not yourself," Moctu said, frowning. "Did you get hit on the head?"

"They drug her," Pele said. "She eat something—drug."

"Beavers," Elka said brightening and breaking her silence. "There were beavers!"

Despite their perilous circumstances, both men snorted and looked down, stifling laughter.

"Yeah, drugged," Moctu agreed. This'll be interesting."

Chapter Seventy-Nine

"SHE BOKE MY TEEF," Devu complained as he held a hand to his bleeding mouth and nose.

"Yeah, her wolf got loose and attacked me," Naramsin said rubbing his own wounds where Misha had savaged him on his right arm and hand.

The Lion leader, Rolf, and the young warrior who had helped Naramsin revive Devu just nodded grimly. It had taken some time and effort to wake Devu and bring him out of his confused jumble of pain and drug-induced visions. He had screamed several times in pain and terror during the ordeal, but much of the effect of the drug had now worn off.

"Sleeps With Wolves must have head-butted you," Naramsin said soothingly to Devu. "Looks like a broken nose and your upper teeth… But we'll fix you up as best we can."

"She need ta die," Devu shrieked, some bloody drool dribbling from the corner of his mouth. "I'll burn her, cut off her bress, rip off her hair. I'll drown her in poson."

Naramsin's head cocked, and his eyes narrowed at Devu's last ranted comment. Knowing Devu was still muddled, Naramsin explored the remark as casually as he could. "So, you'd poison her? What would you use?"

Devu's lip curled exposing his broken front teeth, and he snarled, "Oh, I make her gaad poson, stron poson. She die." He smiled, but it must have hurt, because his hand went to his mouth, and he gently rubbed his bloody upper lip.

"How do you make the poison?" Naramsin asked, and Rolf narrowed his eyes, wondering about all the questions.

"Yew poson, stron poson. She coff, foam come out of mouff, then she die," Devu said allowing himself another smile at the thought.

Naramsin's teeth clenched as he recalled his uncle Tebow dying years before in a similar fashion after arguing with Devu—shortly after a dinner. Tebow was a strong man, and Devu had killed him by just staring at him. At least that was the story that had been told countless times. Tebow had coughed, foamed at the mouth and died.

Was it Devu's power, or Devu's poison? Naramsin wondered. And Devu's claims that he can't die. Naramsin snorted. Well, he was almost killed tonight—Sleeps With Wolves and the wolf sure proved that. But Tebow, he was a good man. Like a second father—taught me most of what I know about hunting. Naramsin's lips tightened, and his eyes grew hard. He found he was glaring at Devu, and he had to look away to control his anger.

Unaware, Devu ranted on. "I wan her caught. Send all of dem, all our warriors. Catch her."

"We need to keep some warriors here for protection, and some of the men need to hunt, Rolf countered,

"No," Devu bellowed, his eyes wide and the veins showing on his forehead. He caught a breath and seemed to calm somewhat. "Okay, keep some fo protectcun, but send da res."

When Rolf stroked his beard to consider the matter, Devu again raised his voice. "Now."

When Rolf nodded deferentially and quickly left the hearth, Naramsin blinked in surprise.

"I always thought they were equals," muttered under his breath. "But crap, has it always been like this?"

Chapter Eighty

"We've got to move faster," Moctu said, picking up the pace. "They know where we're going, and they won't need to search for our trail. Their warriors started out today well-fed and rested, and they'll move faster than we can. I'm just glad we walked all night, or they'd have us by now."

"Sorry I slowed us down last night," Elka said sheepishly as she walked close to Pele. "It seemed so important—what I was thinking about. But..." Her voice trailed off.

Moctu nodded. "But now you can't remember what was so important. Yeah, that's...sort of how it works most of the time." After a pause, he grinned and added, "You'll have to tell us more about those beavers sometime."

"How far...Etseh?" Pele asked, looking at their back trail.

"Three days. Two if we walk at night too," Moctu said. "I'm just worried that they have so many more warriors than we have back at Etseh, and some of ours are still injured from fighting the Sea..." He looked at Pele. "Well, some of your people."

Pele had not been surprised when told of Zeke's treachery in killing the Sea People leader, Arno, and then leading the group into battles with Moctu and the Nereans.

"Yes, Zeke—he...um, no can trust," Pele said shaking his head. "Bad man."

"Well, he's gone now," Moctu said. "And two of *your* guys shot him. Otherwise, I'd be dead."

The small group paused as they came to where the forest met a broad, burned-out stretch of terrain. To the north, the burned-out area was even more expansive, and they stared at it somberly for a long moment.

Moctu shook his head. "We'll cross this and stay in the wooded parts. We could walk faster in the burned areas, but if we stay out of sight in the brush and forests, they at least have to worry a little that we'll ambush them."

"With one bow between us?" Elka asked.

"They don't know for sure what we have," Moctu said. "That's our only advantage. Plus, maybe we can find something to eat. I'm starving."

The group crossed the bleak, ruined landscape in gloomy silence. An empty numbness came over Moctu, and he was muttering to himself by the time they reentered a brushy area with small trees.

"What are you thinking Father," Elka asked.

"Well, even if we make it back, they can likely beat us with so many warriors. And even if we manage to fight them off, we don't have enough food for the winter—which is already here. And the Lion warriors have bows now too. Why'd you show them how to make bows Pele?"

"I took all his bows," Elka said, defending him. "He didn't have anything to trade."

"But," Pele said, then rubbed his forehead searching for the words. "I show them wrong way."

"What?" Elka and Moctu said in unison.

Pele grimaced, then smiled roguishly. "I show them wrong wood, wrong way make bows. Their bows not so good. Not shoot far."

Moctu's head tilted, and a smile slowly built on his face. "Like what?" he asked.

Pele nodded. "I show."

For the rest of the day as they hurried through clumps of forest, Pele would point to most tree types and with his other hand pointed down, wave it side to side, indicating the wood was mediocre at best. For the few hickory, yew and walnut trees they encountered, he'd kiss his hand then touch the trees showing they were the preferred wood types to use.

The sun came out, and the wind died down, but it was still so cold that they could see their breath as they crested a low knoll and rested. Pele took the opportunity to show them the bow he'd confiscated off the downed Lion warrior.

"Look," Pele said digging his thumbnail into the wood and making an arcuate impression. "This ver bad. Not hard and not dry." Then he traced the grain of the wood which ran along the length of the bow for a little more than half of it. "Need this…" Pele traced his finger along the grain for the entire length of the bow. "All bow."

"The wood grain should run the full length of the bow," Moctu interpreted, and Pele pointed at him and nodded vigorously.

"Yes, yes."

"So, the Lion warriors all have bows, but they're not very good," Moctu said brightening.

Pele nodded and smiled broadly, happy he was understood.

"Uh oh," Elka said pointing to the east.

Crossing the burned-out area, they could see Lion warriors—lots of Lion warriors, and they were moving fast.

Chapter Eighty-One

"How many do you figure?" Moctu said to Elka, who had the better eyes.

"At least five hands, but some were probably already across the burn and into the forest before I started counting," Elka said, her mouth suddenly dry.

Moctu winced and shook his head. "And they all had bows. I don't care if they're lousy bows," he said. "That's more than we can handle." He looked down and rubbed his forehead and was quiet for a long moment. Moctu blew out a breath and said, "Way more than we can handle."

"We need to move faster," Elka said. "Need to warn everyone back at Etseh that they're coming, and right now, we need to worry about ourselves or they'll catch us."

With Misha trotting happily, but with a limping gait, alongside Elka, the threesome started jogging west, away from the advancing warriors and toward the relative safety of Etseh. Yesterday, it had been a drug-addled Elka that slowed them. Today, it was Moctu who couldn't keep up.

"I'm sorry," he said, hunched over, his hands on his knees, trying to catch his breath. "I just..." Moctu paused, gasping and looking drained. "Chest hurts."

"That's your broken ribs," Elka said patting her father's back. "You're still recovering from your wounds, and you haven't eaten much in days. And we walked all night. We're all—"

"The warriors...they closer," Pele said, pointing behind them.

After a quick glance backward, Moctu resumed jogging, and the others joined in, watching him apprehensively.

As he ran, Moctu searched his mind for any strategies, any answers as to how to deal with this impending crisis for his people. But none came, only more questions.

He shook his head. "Am I leading this horrible pack of hyenas right to my people?"

Exhausted, the idea of stopping kept coming to him. "I'll stay here with the bow and let Elka and Pele go on. I can pick off a few of their warriors and slow down their filthy swarm. Give my people a chance anyway."

But when he slowed and gasped out his proposed strategy, neither Elka nor Pele would have any of it.

"You're coming with us, or we're all staying and fighting," Elka said firmly, knowing her father would never go for that. "When we get to the top of that next knoll, we'll rest a little. Come on," she encouraged, putting a hand under Moctu's arm and helping him along. Pele did likewise on the other side, and they continued jogging up the low slope.

"Feel like a feeble old man," Moctu said embarrassed. But he found that the support did help.

"You're just not as strong as you normally are," Elka said reassuringly. "We're going to make it."

It seemed like forever to Moctu, but the group made it to the top of the low, brushy hill, and they looked back only to see that the warrior pack had again narrowed the distance between them.

"Crap," Moctu said, and even Elka's optimistic resolve began to crumble.

"Oh shit," she said weakly.

"Uh oh," Pele said pointing the other direction. "They get around us."

Moctu's mouth fell open, and Elka put a hand to her mouth. There was another party of warriors coming at them from that direction. And they all had bows.

Chapter Eighty-Two

FOR A LONG MOMENT, Moctu looked down at the ground and shook his head.

Pointless. It had all been pointless, and now they'd all die. That thought sent a momentary surge of panic through him—mostly for Elka and his unborn grandchild, but also for himself. But the panic was soon replaced by an empty desolation that utterly swamped him. With Pele's help, he'd rescued Elka only to ultimately fail? The thought of never seeing his first grandchild gave him an ache, a hollowness that he'd never experienced before.

It was then that Moctu remembered his father's counsel: "You live your best life, and die your best death. That's all you can do sometimes." The words calmed him, and he immediately bent to the base of a long, stout sapling, and using the knife that Pele had earlier given him, he began cutting it down.

His teeth clenched, Moctu growled, "They're not going to take me without a fight." After a pause, he shook his head. "That smaller party must have taken the southern route, and they ran all night and all day to get around us. They're going to be tired."

Thinking out loud and rather businesslike, Moctu summarized their options as he saw them. As much to himself as to the others, he said, "Can't go to the north because there's so much burn-out area—no place to hide. Can't go east—that's for sure. That's the biggest group of warriors. Now, can't go west and run into the smaller group. That leaves going to the south or trying to fight our way past the small group."

The sapling down, Moctu quickly stripped the branches off and began shaping a sharp point.

"What do you two favor doing?" Moctu said looking up, his face grim but determined.

"South, we go south," Pele answered. "And fast. Big group warriors getting closer."

Elka was quiet, staring toward the west, her hands cupped over her eyes, blocking out the sun.

"West," Elka mumbled finally. Then she snorted loudly, and Moctu looked at her to see if it was in fear. She snorted again, and this time, it sounded more like a laugh. Elka turned to Moctu, and with her blue eyes sparkling, she smiled broadly. "West," she said again. "We're going west. That's Nuri down there. And Palo…and some others."

Tears began rolling down Elka's cheeks which looked incongruous as she chortled with laughter.

"Nuri?" Moctu asked in disbelief, shielding his eyes to look. They were still far away, but Moctu could make out Nuri, Palo, Nabu and two others.

"Spirits, thank you," he whispered, blowing out a breath that turned into a laugh of his own. He turned and grabbed Elka in a bear hug, both of them bouncing with joy.

They broke apart when Pele said with urgency, "Need go. Fast now. They come."

As they ran westward toward Nuri's group, Moctu's smile faded, and he chewed the inside of his cheek. There was still so much to worry about. But as they neared the small group of warriors—his warriors, his people—he allowed a tight smile to return. Moctu knew his people had nowhere near enough warriors to stop the horde coming from the east, but at least for now, they would live on. There was hope. And he would see and hold Nuri one more time.

And there she was. Moctu and Nuri ran to each other, and he swallowed her in a warm embrace.

"I was so worried," Nuri said. "I thought…I thought you were…" She began to tremble.

"I know, I know," Moctu said hugging her tighter and wincing a little at the pain in his ribs. "Going there was…stupid. But—"

"Well, at least it worked," Nuri said, smiling and pushing back to look at his face. Her eyes were brimming with tears as Palo and Nabu came up to pat Moctu on the back.

"You brought Elka back with you," Nuri said breaking away to hug her daughter. "And this must be…"

"This is Pele everybody," Elka said. "He's my…" She looked at Pele questioningly, and he smiled back at her. "He's…my…uh, he's…mine."

Nuri tilted her head at the news, knowing some big things had changed, and that Pele must not have been involved in Meko's death after all. Otherwise, Moctu or Elka would have killed him. Nuri formally, but somewhat coldly, gave Pele a quick hug as he was also greeted by the others.

"I'll need more details soon," Nuri whispered to Elka as she winked at her.

"And really good to have you back," Palo said loudly to Moctu while smiling broadly. "But did you…" He pointed to the Lion warriors that were just cresting the hill. "Did you have to bring those guys with you?"

"Yeah, we'd better get moving," Moctu said to the group.

As they ran toward Etseh, which was about a day away, Nuri stayed close to Moctu.

"Where's Koby?" Moctu asked about his son.

"He's safe," Nuri replied. "Back at what's going to be our last position, if it comes to that." She slowed a little and got even closer to Moctu.

"So, I thought this Pele guy had betrayed Elka," Nuri whispered. "I thought he was one of the two that…"

"Turns out she was wrong about that. He's…he's not a bad guy," Moctu said. His brow wrinkled, and in a hushed tone he continued, "But we've got these other guys to worry about now. Did you see the numbers of those warriors back there? How're we ever going to stop that group? They'll swamp us."

"We've been working on that," Nuri said smiling enigmatically.

Chapter Eighty-Three

"Aren't we going back to Etseh?" Moctu asked, his ribs hurting and his exhalations fogging in the cold air.

Underfed for days, and still troubled by his earlier injuries, Moctu was having trouble keeping up with the rest of the group. Each time he gasped for air it was like a burning log rammed into his side. And it was going to get worse before it got better as Nuri directed the group on a path to the north and urged them to speed up.

"We'll get there," Nuri replied, "It's not that far now. We're set up about a half day's walk from Etseh. But we need them to follow us on the northeast pathway there."

"Oh, so you have some…" Moctu sucked in another breath, laboring not to slow the group down. "Some pit traps set up along this way." He smiled knowingly. "But that's…that's not going to stop this swarm. There's so many of them."

"We have other…" Nuri paused and smiled the same enigmatic smile as before. "Lots of surprises waiting for them."

"Are Lehoy and the others setting up an ambush—something like that?"

Nuri nodded. "Something like that."

Moctu studied her for a long moment as they ran, then walked, then ran again. He was extremely proud of her. It was clear that she was in charge, directing things. And she was calm.

"How can you be so calm?" he whispered, not wanting the others to hear. "We have five or six hands of Lion warriors…they're bearing down on us—and they're all armed with bows." Moctu sucked in another breath and shook his head, frowning. "We just don't have the numbers to stop them. It'll be a bloodbath. I'm willing to die, but I'm not calm."

"I'm not calm either," Nuri whispered back. "But with our… surprises, I think we have a chance. But right now, we have to pick up the pace. Can you make it? I think I can even hear them behind us. This all falls apart if they catch us."

The small group maintained the walk-run-walk routine during the course of the afternoon, passing through mostly low, brushy terrain with tall sedge. They'd gotten frequent glimpses of the huge band of warriors trailing them. The Lion warriors weren't far behind, but they hadn't narrowed the distance either. As darkness fell, the small band was exhausted—especially Moctu.

"Sorry, no stopping," Nuri advised the group. "But it's best to walk so we don't trip or twist ankles. We'll run again if we hear them getting close."

Nuri gave Moctu her last piece of deer jerky, and it briefly revived him, but only enough to let him worry more. Moctu tried to shake it off, remembering the guidance his stepfather, Tabar, had given him: "When you're tired, you're brought down by ideas and worries that you conquered long ago."

But it was no use. His people were threatened with slaughter, and he couldn't figure a way to stop it. Worries gradually succumbed to exhaustion, however, and by midway through the night, Moctu was overcome by a dull lassitude. Through a bleary, mental haze, he could barely find the wherewithal to put one foot in front of the other. Relief touched with embarrassment spread through him when Palo slipped alongside and carefully draped Moctu's arm around his shoulder to help support him.

"Thanks Palo," Moctu said. "That helps. My ribs…they're…" Instead of finishing the sentence, Moctu just blew out a loud breath.

It was a clear, bitterly cold night, and the moon was still nearly full. The Moon Spirit, a woman, had begun her menses, and she'd started to get smaller. With each moon cycle, she shrank to nothing and then was reborn. Nuri stared at the bright orb for a moment and allowed herself a tight smile. At least its fullness hadn't marked the end of Elka or Moctu. For that, she was thankful to all the Spirits. But now, there was this new, bigger threat—bigger because all her people were endangered.

On a hillock, Nuri looked back and could see the lights of three fires indicating that the Lion warriors had made camp. She eyed Moctu, knowing that, even with Palo's help, he was struggling with his wounds, and from malnutrition and exhaustion.

Nuri took his other arm and said, "They've stopped for the night, so we can go a little slower. How are you holding up?"

Coming out of his stupor, Moctu's eyes cleared, and he smiled abashedly. He looked to Palo, then to Nuri.

"Can't believe it's come to this—where I'm the one holding everyone else up. It's humiliating. I'm sorry."

"Oh, come on," Nuri and Palo said in unison. "You're doing great," Nuri continued. "You've been in two horrible fights, and you've got wounds, serious wounds. And then lately, you've been mistreated—starved and frozen for days. I'm amazed you've been keeping up. We're doing fine. We're getting close to where we're set up."

Palo peeled off, and the couple walked together in comfortable silence until Nuri smiled and nudged him. "Look at them," she said nodding her head toward Elka and Pele, who walked side by side, his arm firmly around her shoulder. Misha devotedly walked close by on the other side of Elka.

"They look like they're in love," Nuri said.

"Yeah, I think so," Moctu said. "Quite a threesome there."

As dawn broke, frost glistened on the sedge, and the robins, blackbirds and thrushes began their morning songs from the more forested landscape the group was now entering. It was only a half day's walk to Etseh from here, and as the path narrowed, Nuri put her hand up to slow the group.

"We're almost to where we're set up. We have to be really careful from here on," she cautioned. "Lots of pit traps. Palo and I know where most of them are, but Leuna and the others were building more as we left, so there's some others—we'll just have to go slowly. Walk where I walk, but don't leave a trail that the Lion warriors can use to avoid the traps. Elka, keep Misha close to you. Nabu and Palo, can you please follow us, and use a branch to brush out our tracks and sprinkle some dry leaves over everything?"

The group had only gone a few dozen steps into the dense forest when Misha turned, snarled, then barked out a sharp warning. That's when the Lion warriors came shrieking at them from behind.

Chapter Eighty-Four

"They followed us last night," Nuri cried. "Those campfires were a trick. Hurry now. Follow my footsteps—fast!" With that, Nuri dispensed with her careful, disciplined maneuvering, and she rapidly wove her way through the network of pit traps.

"Watch out for that one," she yelped as her heel almost went into a newly laid trap.

"Kill them. Shoot them!" screamed several Lion warriors not far behind them.

Arrows began to whip past the group but none hit, and fortunately, the trail curved, blocking further shots.

"There!" Nuri shouted, and she pointed ahead to where several Nereans waved their arms from behind a large deadfall.

Moctu could see Lehoy, Seetu and Alfer, and there were perhaps a couple of others moving, but mostly concealed, behind the huge, fallen pine tree.

"I hope they brought some extra bows for us," Elka said as they sprinted toward the natural bulwark.

"Aaeeii!" someone screamed in pain from behind them.

"Got one," Nuri said loudly, happy that one of her pit traps had found a victim. "That'll slow them down some."

"Still outnumber us more than two to one," Moctu muttered anxiously. "We'll never be able to hold them off. They'll flank us and surround—"

"Got a plan for that," Nuri cut him off.

The fleeing group reached the battlement and skirted to the right around the massive root ball that had been pulled up when the tree fell. Pele bent to avoid one of the scraggily roots that splayed out from the downed tree just as an arrow whizzed above him through space he'd occupied a heartbeat before.

"That was close," Elka cried. "Hurry!"

But another arrow slammed into his lower leg with a sickening whump, toppling him.

"Aaagh," he bellowed in pain as he fell into the depression left by the upturned root ball.

"Pele," Elka screamed, as she raced to pull him to safety with Misha bounding along by her side. Arrows whipped past Elka as she struggled to pull Pele from the crater. Another grazed Misha causing her to yelp and run to the side.

From their positions behind the tree trunk, Lehoy and Seetu provided covering fire, and Lehoy scored a hit on the leading Lion warrior. It was his second shot, the arrow taking the burly warrior in the stomach just as the man released his shot which flew wildly high. Palo rushed to help Elka pull Pele to safety, and the hail of arrows slowed.

The arrow had pierced the front and back of Pele's leg and carried through almost to the feathers.

"Lucky—it hit the side of your leg, not the thick part," Palo said matter-of-factly. "Came all the way through. Happened to me once. Here, bite this."

He gave Pele a wad of leather to bite down on, and quickly broke the arrow on the back side of his leg where it was longer.

Pele closed his eyes and twisted his head in pain.

"Aaagh!"

"That hurt, but this'll be worse. Gotta pull the rest out the other way cause of the feathers," Palo said, and he immediately pulled hard on the broken piece still in Pele. There was a sucking sound, and the arrow came out along with a gush of blood that rapidly lessened to a steady stream that dribbled down Pele's leg on both sides.

"Ooarrgh!" Pele groaned, grimacing and biting on the leather.

"Okay, it's out, and all that bleeding will help flush out any poison. And you're lucky it didn't hit a big vein," Palo said.

Pele opened his eyes and, still in great pain, gave a curt nod of thanks for the aid. Palo nodded back and continued, "If you can stand, we need you shooting. Elka's told us how good you are with a bow."

Pele spit the wad of leather out and tried to get up.

"Yes, I shoot," he said, struggling to stand using his one good leg.

Reluctantly, Elka helped him to his feet, and Palo smiled, nodded once more to Pele, then nocked an arrow and peered over the log.

Elka retrieved Pele's bow, and gave it to him, and looking around, saw that Alfer had two extra bows.

Crouching, she ran to get one, and Alfer pointed to where two quivers of arrows lay.

Elka recognized the closest bow as one of the four she'd long ago taken from Pele, so she grabbed it and ran back to the battlement. She looked for a target while she strung the bow, and that's when she saw him, or at least his red tunic. Through the low forest undergrowth Elka saw a flash of red and enough features to be sure the man was Bullow. The man who killed her brother and who nearly killed her was leading a group of warriors around to the left.

Elka nocked an arrow, but Bullow and the group stayed behind trees and scrub, safely out of bowshot.

"Come on, Bullow, I want a shot," Elka murmured, her teeth clenched tightly.

"Aaowwww!" came a cry from behind Bullow's group, and Nuri thrust a fist upward and cheered.

"Traps got another one," she said smiling.

But Moctu frowned and yelled, "On the right," and he pointed to where a group of warriors were moving to flank their position behind the fallen tree trunk.

"They're coming around us on the left too!" Nabo yelled, and Nuri shook her head grimly. Pele fired an arrow at the warriors on the right, hitting one in the arm and briefly scattering the rest as they ducked for cover.

But by moving a little farther back into the woods, the warriors safely reformed and continued their flanking maneuver.

Nuri reluctantly motioned for the group to move back to avoid being surrounded. For the first time, she looked very worried.

"Crap, I thought we'd hold them here longer and hurt them worse," Nuri said tensely. Her brow furrowed, and she begrudgingly said, "Move back—follow me. There're still some traps behind us."

"We got a few of them, but they got some of us, too," Moctu whispered to Nuri as he walked behind her. "Can't keep trading wounds like this when they outnumber us two to one. They're winnowing us down to nothing."

"I know," Nuri said in a strained voice. Then, a bit calmer, she repeated, "I know."

Chapter Eighty-Five

Dusk was nearing as the small group of tired warriors stopped to rest as they crossed another of finger of terrain that had been burned. Moctu pointed to the north where the burned landscape broadened and ran to the farthest hillside.

"That's not from a lightning fire. The Lion people did that. Maybe not all of that on purpose, but…" He shook his head. "They're burning forests and scrublands to drive out game. Devu's got them believing that's the best way to hunt. And for a while it worked great for them, and they killed a lot of game. Lots of times, way more than they could use. But sometimes, the fires get out of control. And now…" Moctu gave a short, bitter laugh. "Now, the game avoids what little forest is left."

Nuri's eyes widened, and her hand went to her mouth as she surveyed the ruined countryside to the north.

"So that's why… Oh, Spirits, that's why our hunts have been so poor."

"Yeah, over these past years they were getting more meat than they could use, while we were getting less and less," Moctu said scowling. "And now, we're both not getting enough."

"Are you kidding me?" Palo said. "Spirits, didn't they realize…" He spit to the side in disgust.

There was a scream from some distance behind them, and Nuri's smile returned.

"Traps got another one of them," she said. "But they're too close. Gotta move on."

The group was silent as they crossed the burned out stretch and reentered the thick forest where it was darker and colder.

"Where to now?" Moctu asked.

"Our last fortified position," Nuri said frowning, then biting a thumbnail. "We have one more surprise for them, but…"

"Yeah, they still have maybe five hands of warriors to us… What do we have?" Moctu looked backwards at their group. "We have about two hands. Those are tough odds."

They walked dejectedly for a long moment before Moctu tried to dredge up some optimism.

"Well, Pele says we have better bows than they do, and we're probably better with them too," he started. "And defending a place is easier than attacking. Some of their warriors probably had to help the wounded get back to their shelter and healers, so maybe…" His voice faltered as he ran out of any positives.

Appreciative of his optimism, Nuri brightened, but it didn't last. She cleared her throat.

"Didn't want to tell you this, but we're out of meat at Etseh. I mean everything. No caribou, no deer—nothing. Down to eating soaked and cooked acorns mostly." After a few more steps, she whispered, "Moctu, what are we going to do? Even if we survive these Lion cockroaches, we're going into a winter with no food." Her eyes welled with tears as her brave front crumbled. "A half moon ago, we managed to deliver Avi's baby girl, and all I could think about afterwards was how's that baby going to make it through a starving winter?"

"One thing at a time," Moctu said, his jaw set. "Let's get to that defensive position…that last position."

Chapter Eighty-Six

THE MORNING WAS COLD and bright, and Moctu nodded appreciatively. Nuri had chosen wisely. After warm greetings from Koby and a few others, the small group of Nereans were now settled into a well-fortified position about a half day from Etseh. They were on a hillside in a rocky bastion with an apron of dense woods below. Rocks had been stacked and piled to form a stout defensive wall, so they had cover and a good field of fire below and to the sides. The group was close to the path to Etseh, so if the Lion warriors went that way, the Nereans could swing down and hit them from behind.

Even so, the mood of the group was grim. The many campfires in the trees below were a constant reminder of the odds they faced, and they saw enough movement to be sure the Lion warriors were still there in force. The small Nerean group had almost no food, and Alfer had begun focusing his gaze on Misha and loudly rattling on about how good wolf meat was—only partly in jest.

With Misha curled up comfortably beside her, Elka scowled at Alfer's latest comment, then turned and asked Nuri, "What's taking them so long? Why are they waiting down there?"

"Maybe licking their wounds," Nuri said beginning to chew on the thumbnail again. "They'll come soon enough." Wanting to change the subject, she said, "Tell me about Pele." Nuri watched as Elka countenance immediately softened, and her eyes sparkled.

"Oh, he's wonderful." Elka shrugged. "He saved my life. Saved Misha's. Rescued us from Devu. I'd do anything for Pele. I was…so wrong about him. It was Bullow who attacked us and killed Meko. He's the one in the red tunic. I'll be watching for him."

"So, you love Pele?" Nuri asked, her eyes narrowing.

Elka shrugged, then nodded. "Yeah. I'd…I'd do anything for him, and I think he loves me too."

"And you do have his baby in there," Nuri said smiling and touching Elka's stomach.

Elka patted her belly with satisfaction. "Uh huh, I was so worried that no one would ever want me, and now…"

"How about your…the sadness times? Are they…"

"They're gone," Elka said brightening even more. "They're…ever since I met Pele really. They're just…I mean, even when I thought Pele was the killer they didn't come back. They're just…gone."

"So happy to hear that," Nuri said smiling. "May have something to do with that baby inside you."

"Just knowing that someone wanted me… A man, I mean. And a really good man." Elka's posture straightened, and she beamed. "He *is* a really good man."

Nuri hugged Elka, and she maintained a tight smile even as her thoughts shifted to how they were going to keep all the new babies and their people alive over the winter. And that was if they survived the Lion warriors.

"They're coming," Palo yelled, and he and others quickly strung their bows and readied arrows. "There's four, maybe five hands of them at the edge of the forest."

From the rocky stronghold it was a long bowshot across a sloping, rocky meadow to the wood's edge.

An unseen man yelled from behind a thick oak at the wood's edge, "Moctu, give us Sleeps With Wolves now and we won't attack. You can live. Otherwise, we're going to kill you and all your men and take your women. There won't be any more Nereans."

Elka's face clouded and she looked conflicted, not wanting others to die or be injured because of her. But Pele put a reassuring hand on her shoulder, smiled and shook his head.

"Just come and try," Moctu yelled back, "…and we'll leave your bodies to rot on this hillside."

"No need for a lot of your people to be killed," the Lion warrior yelled, stepping from behind the tree. "Just give us Sleeps With Wolves, and none of you have to die here today."

"That's Bullow," Elka said excitedly, recognizing the man's red tunic. With a snarl, she hurriedly readied an arrow on her bowstring and called to Nuri, "He's the one who killed Meko."

"Okay," Bullow continued. "You want it the hard, bloody way, that's what you'll get. We outnumber you by a lot, and we have bows, lots of bows."

There was a whirring sound, and an arrow thudded into the oak, barely missing Bullow, causing him to yelp and dive for cover.

"No Elka. No Sleeps With Wolves," Pele yelled.

Moctu had been eying Pele after the 'lots of bows' comment to see his reaction. The big man had taught the Lion warriors how to make and use bows. But Pele had now traded out the crude, Lion warrior bow he'd taken for one of the good ones that the Sea People had made. Even with a wounded leg, Pele had almost hit Bullow with his first shot.

"Okay then," Bullow yelled regaining some composure, but staying mostly behind the oak. "You want it the hard way, that's what you'll get. We'll kill all the men and keep all the uninjured women for ourselves." To his men he yelled, "You want women? Well, go get 'em."

There was a muted cheer from his men, and from the cover of trees, most of the Lion warriors began sending volleys of arrows into the narrow rocky redoubt while others zigzagged their way up the hill. Arrows thudded into the ground behind the Nereans or clattered off rocks. A glancing arrow took Nabu in the shoulder, embedding itself about half a finger length in where it was stopped by bone.

"Crap," was all that Nabu said, but he dropped his bow and sat dazed on a low, flat rock. Nuri quickly put down her bow and went to attend him.

"It doesn't look like it's stuck too far into the bone," she said, "So let's get it out now."

Nabu bit down on the side of an arrow, and Nuri placed her knee on Nabu's side, grabbing the shaft with both hands.

"This'll hurt—bite down hard."

"Aaaagh," Nabu gasped as she pulled the arrow out, and Nuri smiled, happy that the whole point had come out and not broken off in the bone.

"Got one," Moctu shouted, clapping Pele on the back as a Lion warrior limped back toward the forest with an arrow in his hip.

"Me too," Palo yelled.

"I did too," said Lehoy happily just before an arrow caromed off a boulder and hit him above the right collarbone. The arrow had lost much of its force, but the tip still penetrated all the way through the muscle and protruded a finger length out of his back.

"Ooooooh," he said and sat on the ground.

First one, and then all of the advancing Lion warriors turned and moved quickly downhill helping their wounded comrades.

"That's another one," Palo yelled exultantly as he watched an arrow fired by Pele hit a man in the backside. The warrior had been helping one of the wounded, and now the two struggled to help each other.

"And they're retreating!"

Palo turned, and his excited smile swiftly faded as he saw that Lehoy and Nabu had both been wounded.

Nuri was oddly quiet and immobile—except for chewing on the remnant of her thumbnail—while she stared fixedly at the retreating Lion warriors.

Moctu started toward Lehoy, then stopped to study Nuri.

"She's more worried than I am," he murmured, feeling an ache in his heart. "Just better at hiding it."

Chapter Eighty-Seven

The Lion warriors came no more that morning, but there was considerable activity in their camp and the sounds of hacking on wood. It was dusk when they attacked, and this time, they came up the hill behind two massive shields.

"This is trouble," Palo said loudly as arrows began to clatter off the rocks of their fortification.

The shields were made of dozens of saplings the length of about two men. Thick shafts had been laced together and hung down nearly the height of a man. Two strong warriors carried each shield shoulder high while other warriors crept closely behind the protective barrier, their bows ready for action.

"Crap, there're about three hands of them coming and there's... at least another hand firing at us from the woods," Alfer yelled, his voice an octave higher than usual.

"And Lehoy and Nabu can't use their bows," Seetu cried. "That only leaves about a hand of us—do we retreat?"

"I can still use a spear," Lehoy said grimly, hefting one with his good arm.

"Me too," Nabu agreed.

Misha growled and readied herself for a charge, but Elka motioned her to stay. "Soon enough, Misha, but let them get closer."

Moctu fired an arrow that thudded into the top sapling of the nearest shield just below the neck of one of the shield carriers. The man instinctively ducked, and the shield dipped momentarily, but it was up again in an instant and carried a little higher now.

"They're going to be on us soon," Seetu shouted.

"Aim for the shield bearers," Moctu yelled. "If a shield goes down, we've got 'em."

Arrows began thudding into the shields, but none scored a hit, as the attackers edged ever closer.

Nuri had not fired an arrow. Instead, she stared off at nothing, her lips pursed. When the shields and the Lion warriors behind them were more than halfway up the hill, she finally nodded to herself, and she let out two odd, ululating and piercing whistles. As her group looked at her questioningly, she calmly sighted her bow and fired.

The shield bearers made it several steps farther before the commotion started below them. Bushes moved behind the tree line, and spears emerged, cutting at the Lion warriors still on the forest edge.

"Oh, crap," Moctu yelled. "It's the Shiv."

"What the…?" Palo said loudly.

Moctu flashed on wisdom his stepfather, Tabar, had shared long ago, "The enemy of my enemy is my friend—for a while."

The advance of the shields faltered as the bearers turned to see what was causing the pandemonium below. The leftmost bearer let his shield drop too much, and Pele's arrow took him in the lower throat, falling him instantly. As he fell, the whole shield came tumbling down, and Palo dropped another warrior with an arrow to the chest.

"Shield's down—shoot 'em," Alfer screamed.

Nuri shouted something that was lost in the clamor.

Behind the other shield, a Lion warrior wearing a red tunic turned to shoot downhill at one of the Shiv. As he rose a little too high, Elka's arrow took him in the upper back.

"That was Bullow," she shouted. "He's down!"

Pele hit another warrior who had turned to run. He didn't fall, but with an arrow protruding from his rear, he could only limp downhill. Seetu dispatched him with an arrow that hit him in the back of the head. The arrow penetrated below the skull and projected through the mouth, knocking out teeth. The warrior turned slightly and was dead before he hit the ground.

Nuri's shouts could finally be heard above other yelling.

"Only shoot at the Lion warriors," she screamed over and over.

It was then that one of the leafy, well-camouflaged Shiv warriors at the edge of the forest rammed his spear into a Lion warrior's gut, then looked up the hill. With a black and green painted face, the man smiled. Moctu recognized Ronk's two missing front teeth immediately.

"It's not the Shiv," he yelled excitedly. "Those Pale Ones aren't the Shiv. It's the Krog. It's the *Krog*."

Every Lion warrior on the wood's edge was down, and the Krog began to screech an odd, high-pitched battle cry as they move rapidly uphill. In absolute terror, the remaining warriors behind the shields broke and ran. Moctu now recognized Hawk who smiled and held a fist up toward him in a salute.

"Brud Hawk," Moctu murmured and held a fist up in a solemn return salute. "Spirits, how'd they know? How'd they…?" With a questioning look, Moctu turned toward Nuri who smiled knowingly at him and gave him a crisp nod. She raised her fist in a salute to Hawk as well.

The momentary interlude was broken as Lehoy and Nabu charged over the rock wall with spears in hand roaring and whooping their own battle cry. Moctu and Palo followed with bloodcurdling screams, each still firing arrows at the retreating warriors.

"Now, Misha," Elka cried, and she and the wolf charged over the wall as well.

Perhaps it was the relief of his people no longer facing extermination. Or it may have been the sudden change of fortune from demoralized despair to exhilarating victory. Whatever it was, Moctu felt an overwhelming joy on a level he'd never experienced. They'd crushed the Lion warrior assault, and none of his people were killed. The closest Moctu had been to this euphoric feeling was long ago when he'd helped kill two mammoths—and that had been with Hawk as well. Moctu broke off his chase and turned to look for his friend.

From a distance, Moctu winced as he saw the Krog leader, Da, spear a badly wounded Lion warrior, ending the man's pain. Just to the right of Da, Hawk emerged from the woods and smiled at Moctu, once again holding up a fist of salute. Moctu briefly held his fist up in answer, then broke into a run to embrace his friend.

"Brud Moctu," Hawk called as he neared, and then they were together, hugging and jumping in jubilation.

Chapter Eighty-Eight

Early winter days were short, and dusk turned quickly to darkness. With torches, all available arrows were collected, and the Lion warrior bodies were stripped of anything useful. A rough count indicated nearly three hands of dead warriors. Moctu looked for Naramsin among the dead, hoping he wouldn't find him. But the scar-faced man was not among the corpses, so he'd either escaped or was never there.

The Nerean's wounds were cleaned with fresh water, dressed with sphagnum moss, and wrapped tightly. If any of the wounds were still bleeding heavily in the morning, they'd be cauterized with red hot flint.

The Lion warrior camp was a treasure trove of bows, arrows, furs, and best of all, jerky and other food. A frigid wind began to blow, and it wasn't long before the entire diverse group—except for one guard—was sitting around a giant bonfire, laughing, telling stories about the fight, and eating the Lion warrior food.

Earlier, Nuri had recognized Lok as one of the camouflaged Krog warriors, and she ran to hug the powerful woman.

"Lok, you came," Nuri exclaimed. "Thank you so much! What about little Mac, your one-year-old? He was still nursing."

"Seve, she nurse him," Lok said, and then after a pause, "I come hep you."

"Ooh," was all Nuri could croak out before bursting into grateful tears and hugging her friend tighter.

The two women had shared a horrible time years before as captives of the Shiv, and together, they'd overcome that dreadful situation. They shared a bond that no other women could possibly fathom.

Lok, Hawk, Da and Ronk had all come to help Nuri and Moctu save Elka. Although Hawk and Lok spoke some Nerean, the four Krog ended up crowding around Elka who sat with Pele, because Elka was fluent in both languages and could translate.

Smiling broadly, Ronk soon spouted a long story in Krog that caused the other three Pale Ones to laugh uproariously. Elka translated for the others around the fire.

"They were late getting here because Da and Ronk only got back from a hunting trip a few days ago. Nuri had told Hawk and Lok of this site and how to find it, and the four of them rushed here, but the Lion warriors had already arrived. After making sure their camouflage was perfect, they crept in carefully last night. Ronk says this afternoon they cut down a sapling right next to him, and one warrior almost pissed on him."

The crowd was already in a celebratory mood, and this story convulsed the group. Palo went to his knees roaring with laughter, and that only brought on more laughter from the group. Elka laughed so hard her eyes streamed with tears. Although not understanding much of the story, even Pele laughed because Elka was laughing so hard. Hawk clapped Ronk on the back as the laughter lessened, and Moctu took the opportunity to stand.

"I want to give thanks to the Spirits for this victory," he said. "But also, to all the folks here who fought so bravely." Moctu frowned and looked down before continuing. "It looked so bleak there for a while. We were so heavily outnumbered. But our Krog friends…" Moctu's frown transformed to a grateful smile, and he waved a hand at Hawk and the others, "The flood of Lion warriors was coming at us, and arrows were raining down on us from the tree line, but the Krog turned the tide. They took down every Lion warrior at the tree line then started up for the others. That's when the Lion shield line broke, and when they started to run, we hit them from behind." Looking to the Krog, Moctu put a hand to his heart. "We owe you so much—thank you. *Thank* you."

As Elka translated to the Krog, cheers went up from the Nereans, and Lehoy and several others nodded appreciatively to the Krog. The Krog looked embarrassed but pleased.

Moctu continued, "So, thanks to the Spirits, and thanks to the Krog, and thanks to all who fought so bravely." When the cheering died down, Moctu looked toward Pele and said, "And Elka wouldn't be here today...well, neither would I. *Neither* of us would be here if not for our new friend, Pele." Moctu pointed to Pele and put his other hand to his heart. "Thank you, Pele." Moctu glanced at Elka and smiled mischievously, then looked back to Pele and said, "We all hope you'll stay around here for a while—a long while."

Elka blushed as people chuckled.

"Finally," Moctu started again. He turned solemnly toward Nuri. "One more...there's one other person deserving great honor. And it's my mate, Nuri."

Moctu's eyes met hers, and he could see tears welling up in them, as others in the crowd murmured agreement.

"I've always been proud of you, but never like this. You planned this. Got all the traps and defensive positions built. Visited the Krog and got them involved. Planned the encounters and fallback positions. It's amazing, it's just..."

A tear tracked down Nuri's cheek as she stood to embrace Moctu, and the whole crowd cheered heartily.

"I just thought the Krog should know about Elka, and...and they wanted to help," she said quietly. "I just thought..." Nuri shook her head and finished simply. "I love you."

Chapter Eighty-Nine

EVEN AS THE NIGHT grew colder, the celebration continued. Logs were heaped on the fire, and people had to turn occasionally as their backs were cold, but their fronts were scorching.

Addressing Elka and Moctu, Hawk said, "We come help Brud Moctu and Elka—is *obvious*."

Moctu rolled his eyes and chuckled at Hawk's use of the word, which Moctu had taught him years before. It had rapidly become Hawk's favorite, and he used it over and over.

Hawk continued. "Rah, she want come and help Elka, but…" He grimaced and hunched his shoulders while crooking his hands, indicating that her arthritis prevented it.

Hawk looked questioningly at Elka, and said a string of Krog words which she translated, blushing a little.

"Hawk says he looks forward to telling my Grandmother Rah about this big victory, and that I'm safe—and have a baby coming."

Elka dipped her head as people cheered and clapped, but thereafter, the crowd gradually grew quieter as the adrenaline and elation wore down, and exhaustion set in. People turned reflective, staring trancelike into the fire's glowing embers. More furs were brought out as the fire died down, and friends and lovers huddled amicably together.

Pele's leg wound throbbed, but he smiled contentedly as Elka rested her head on his chest. He touched her slightly protruding belly and said, "My baby. This ver happy. You be my mate, yes?"

A drowsy Elka's eyes opened wider, then closed as a satisfied, happy glow brought a peaceful smile to her face.

"Yes," she murmured. "Yes, of course. That makes me very happy."

Moctu and Nuri got up to move to their sleeping furs, but Moctu first checked with Seetu, the guard, who reported no movements or anything suspicious in any direction.

Moctu yawned, and told Seetu to wake him for guard duty whenever he got tired.

"Will do—thanks Moctu," Seetu said. "I'm doing fine for now.

With that, Moctu retired into the warm, luxuriant furs with Nuri, and she kissed him warmly and deeply.

"I've missed you so much," she said. "I was so worried."

"I missed you too," Moctu said pressing his body closer to hers. "I'm…I'm still amazed at all you did. You're an incredible woman. And a wonderful leader."

As he thought about the day's outcome, Moctu breathed a contented sigh. But it didn't last long, as his thoughts turned to the huge, remaining threat that his people faced.

"Nuri, what are we going to do about food—meat to get us through the winter? How will we ever—"

"One thing at a time, love," Nuri said pressing into him. "We'll think about that tomorrow. Right now, I have other plans for you."

Chapter Ninety

"Spirits, it's cold today," Moctu said in a muffled voice through the skins covering his mouth and nose. "Can't wait to be close to the fire in our warm hearth."

"Be there soon," Nuri said, "and we can unload all this stuff. There's so much."

"Misha's doing her part," Elka said brightly as she and Palo helped Pele hobble up the low hill. Elka had been surprised and pleased last night in the furs when Pele wanted her despite the pain from his wounded leg.

All except the wounded were carrying a full load—even Misha had baggage strapped to her. Besides the furs they'd found at the Lion warrior camp, most of which were being worn this morning as wraps, there were the clothes, moccasins and necklaces they'd stripped off the bodies of the dead warriors, and there were lots of bows, quivers of arrows and spears.

The Krog had departed at dawn with their share of the valuables, after a long series of warm embraces and heartfelt thank-yous from Moctu, Nuri and Elka. When Moctu asked if the Krog had extra meat stored for the winter, Hawk said, "No, we go now—hunt. Need more, more. Hawk frowned, then looked to Elka and said a fast string of words which she translated.

"Hawk says the birds flew south early, and the oaks dropped a lot of acorns, and both of those are signals that this winter's going to be early and long."

Moctu grimaced and his shoulders dropped, but he forced a smile, nodded and hugged Hawk once more.

"Spirits be with you, Brud Hawk."

Now, on the path back to Etseh, Moctu endeavored to be upbeat, but a suffocating darkness lingered nearby.

“You’re lost in thought again,” Nuri said. “Still thinking about winter and our food?”

“How can I not?” Moctu said, frowning. “I’m supposed to keep my people safe.”

After a pause, he continued, “So, we don’t have to worry about the Lion warriors anymore, and we have lots of furs and bows, but we’re going into a harsh winter with no meat. We ate most of what we found in the Lion warrior camp last night. Which leaves us…” He shook his head instead of finishing the sentence.

“Well, we can finally hunt,” Palo said. “I know they’ll have to be some really good hunts. But I’ll be ready to go in the morning. Maybe we’ll get lucky.” His normally optimistic countenance darkened, and he muttered, “That’ll be tough with all the burned-up territory out there.”

Pele cocked his head, “What they say? No food?”

“Yes, Pele,” she said patting her stomach. “We have very little food…not enough,” She hugged her own shoulders indicating cold weather, and continued, “And winter will be cold and long.”

“Yeah,” Moctu agreed, looking at Palo. “That’s the only thing I can think of—a good hunt. And you’re right, we’ll have to be luckier than we’ve ever been.”

As the group neared Etseh, Pele was still being aided by Elka and Palo. He squeezed their shoulders and motioned with his head toward Moctu. Steering them closer, he said, “Moctu. I um…I hep with food. I hep with food maybe.”

Moctu smiled and pointed to Pele’s leg. “When your leg’s healed, that’ll be a big help Pele. You’re by far the best bowman we have. But that’s going to take some time, and that’s…” He shook his head. “That’s time we don’t have.”

Pele didn’t follow all the words, but he understood the tone and gist of Moctu’s comment. He nodded and waved a hand dismissively.

“Fish,” he said simply. “I know…, catch many fish. And seals. But…need boats.”

Moctu’s head twisted, and he stared at Pele for a long moment, his eyes narrowing.

"Boats?" he asked. Then Moctu nodded, and a slow smile built on his face. He put two fingers up. "We have two boats."

Chapter Ninety-One

"You make good net," Pele said testing a woven net that Nuri and Leuna had made to his specifications. As the muscular man pulled hard on the net, Nuri worried that it wouldn't hold up. But it did, and they all smiled.

The mood at Etseh was upbeat, as Pele had convinced most everyone that they could survive the winter by catching fish—lots of fish, and maybe a seal or two. Alfer was among the few complaining that they were wasting time that could be better spent hunting. He and his grandson Itzal had, the day before, headed north to hunt.

Earlier that morning, Pele nodded approval at the barbed, harpoon spear point that Moctu had carved from mammoth tusk.

"This ver good—work good," Pele said holding up the finished piece. He clapped Moctu on the back and held out three fingers. "Now, three more."

Moctu, Nabu and Lehoy had already each carved two large fish hooks made as Pele advised. Based on the size of the hooks, he was expecting to catch some huge fish.

"Spirits, I hope we're not wasting time," Moctu muttered to the other two carvers while working. "I want this to work… We need it so much. I just hope we're not…" He shook his head. "…chasing after the wind."

Palo was sent to the coast to check on the condition of the two sealskin boats they'd received from the Sea People as part of the peace settlement. He and Nabu had, several moons before, pulled them far uphill and hidden them under a rocky ledge. Then they'd covered them with large piles of thorny brush to keep the hyenas and bears away, but gnawing rats might still have been a problem.

After four days of solid work, during which they all got tired of eating peeled, soaked and roasted acorns made into mush, the nets, harpoons, fish hooks and coils of cord were ready, and a large group of both men and women made their way to the coast, hoping Palo had found the boats in good condition.

It hadn't rained for nearly half a moon, so crossing the ice-cold streams hardly slowed the group, and they made it to the coast in two days. At midday, Palo met the group looking glum, and Moctu looked down at the ground and shook his head in disappointment. Then, Palo's glum look turned to a smile, and he laughed and winked.

"They're great—in good shape. Almost no damage."

A cheer went up, and everyone considered it a good omen.

Moctu wanted to go out immediately, but Pele stopped him.

"More to do," he said, smiling patiently. "And evening better for net fish."

That afternoon was spent attaching cord to the fishhooks and harpoons, getting the boats down to the water and testing them for leaks, and, with Elka's help, getting further instructions from Pele as to when and how the nets should be cast and drawn in. But even Elka, who was good with languages and who'd picked up much of the Sea People language quickly, had trouble understanding the big man.

She grimaced at him and held up her hands saying, "Speak slower. I don't have a clue what you're…" She shook her head. "Something needs to change on the nets?"

Unperturbed, Pele got on his knees and began tying bigger weights to the bottom of the net.

"Ah," Elka said finally understanding. "Pele says the seas are rough, so we need bigger weights for the bottom of the net."

As evening drew near, Pele looked at Moctu, gave him a thumb's up and nodded his head.

"It's time," Moctu said excitedly to the others, and he helped the still-hobbled Pele wade a short way into the cold water and climb into the boat. Moctu placed the flat piece of wood Pele instructed him to bring into the boat before he climbed in as well. Palo and Nabu, carrying the heavy net, entered the other boat easily and managed to paddle through the surf. Moctu's feet were wet, and he was cold, but it was his first time in a boat, and he was full of energy and enthusiasm. His paddling was initially awkward and ineffective, and he found himself sweating. Pele's strokes, however, were expert and powerful, and as Moctu mimicked them, he gradually found a rhythm and technique that blended with Pele's.

Pele pointed to an area where gulls, gannets, and other sea birds were circling in the air and occasionally diving into the water.

"There—that good for nets."

But by the time the two boats arrived at the site, the birds were dispersing, and several casts of the net produced nothing.

Moctu studied Pele's reaction, but aside from a brief frown and a shrug, Pele seemed undismayed. Instead, he pointed to a small island to the south, and he nodded for them to paddle there.

"Maybe seal. Shoot harpoon with bows. Much meat."

"But it'll be dark by the time we get there," Moctu worried.

"No, we go fast," Pele said confidently, and he set a pace of stroking that Moctu had trouble matching. Palo and Nabu were left far behind.

As they neared the island, Pele pointed to a rocky shore where two seals were resting, one medium-sized, the other huge. Pele put his finger to his lips, and he began to dip his paddle into the water more softly. They moved stealthily closer to the seals, and Pele put his paddle away and took out his bow and strung it. He fitted the harpoon-tipped arrow onto his bow and adjusted the attached coil of rope. As he rose onto a knee and pulled back the bowstring, the smaller of the two seals lurched into the water, so Pele shifted his aim to the larger beast and released the arrow. There was a whirring sound as the arrow sped toward the remaining seal, trailing the attached cord. The arrow took the seal in the side, and it made a high-pitched yelp and lumbered into the water while Pele secured the cord to the boat.

That's when the problems started.

Chapter Ninety-Two

"Why do they have to go out so far?" Nuri asked Elka frowning. "Aren't there any fish closer to shore?"

Everyone was sick of acorn mush, and they stared out to sea hoping the boats could bring in a big catch of fish. Finding a new source of food was critical—even their stores of acorns were getting low.

Because Sima, Nindai's Pale One mate, and Elka had the best eyes, they were questioned repeatedly by the group about what was going on in the distance. The sun was directly at them, and the gleam off the water was painful, so Elka squinted tightly and cupped her hands around her eyes to limit the incoming light.

"Palo and Nabu just cast the nets," she said excitedly, and the crowd strained to see the action.

"Spirits, please give us a good catch. *Please*," she whispered to herself.

People were hungry and complaining even more these days about being cursed. And Elka knew they blamed her for the curse.

"What'd they get? What'd they catch?"

Elka stared for a moment longer, then her shoulders sagged and she said, "Nothing. Nets were empty." It was all she could do to keep the resignation from her voice as she said, "But they're gathering the nets up for another cast."

In the muttering that followed, Elka heard the word "cursed" several times. She closed her eyes and took a deep breath. Then she squinted and watched three more castings come up empty. As Palo and Nabu stowed the net and began paddling, Elka shook her head and decided it would be best for her and Pele to leave Etseh.

"That is, if he'll still have me," Elka murmured to herself. "May not want a cursed woman."

Looking out to sea again, Elka saw the boats diverge, with Pele and her father moving to the south, and Palo and Nabu setting up in a new area for some further casts.

"Pele and Father…Moctu are moving to the south. Fast," she said to the group. "They don't have the nets, so maybe they see something by that island."

"It's getting late," Nuri worried. "Be dark soon. They should come in." She bit her thumbnail for several heartbeats, then said, "Leuna, Siduri, will you help me gather some wood? Looks like it's going to be dark before they head back in. A couple of bonfires will help them find their way back here."

The sun was halfway behind the horizon, and the glare was reduced, but Elka and Sima could only see dark silhouettes.

"Pele's got his bow out," Elka said in a loud voice. "He sees something," she said breathlessly. She pulled her hands away from being cupped around her eyes, realizing their trembling was bothering her view. "Using the harpoon with the cord," she continued excitedly.

"Did he get it?"

"What was it? What do you see?"

"He got something," Elka said triumphantly. "Think it's a seal! A big one."

A cheer went up from all around her, and Elka glowed with happiness. But that didn't last long.

"Whatever it is, it's big and it's pulling them hard," Elka said grimacing, her elation now crowded out by concern. "Pulling them out to sea. Pele's motioning to Palo for help."

"Oh no, oh crap, their boat's tipping!"

Chapter Ninety-Three

THE BOAT SURGED FORWARD as the seal swam to the west, away from the coast, pulling the vessel with him.

"Big seal," Pele said in a voice that sounded excited yet apprehensive. He waved for the other boat to come swiftly to help, but the seal was pulling his boat away faster than Palo and Nabu could paddle toward them.

"Craaap! What do we do now?" Moctu asked anxiously as the seal dove deep, and his rear portion of the boat rose a foot higher, almost off the water. The covered, prow of the boat dipped dangerously downward, completely underwater, and a wave swept over and into Pele's compartment. Realizing the craft was likely to be inundated, Moctu grimaced with disappointment offset with relief when he saw Pele reluctantly pull his knife to cut the cord.

But the stern of the boat settled back down into the water as the beast resurfaced and moved to the northwest hauling the craft behind him. Pele motioned for Moctu to use his paddle to help brake the boat's movement, and he waved his arm more aggressively for Palo and Nabu to come faster.

Cold water sloshed through the bottom of the craft, and Moctu, who had previously been sweating from exertion, was now shivering.

Only a sliver of the sun remained above the horizon, and the remnant glowed its golden brilliance surrounded by clouds of red and gray. As the last bright edge of it dipped below the horizon, dusk set in quickly, and Moctu could see the Stars of the Warrior begin to twinkle as the giant beast pulled the boat ever farther from the distant shore.

Moctu heard Pele mumbling, then realized it was a chant.

"He's praying," Moctu murmured growing even more nervous and deciding to pray himself.

"Sea Spirit, we honor you and ask for your blessing. Please allow us this heroic creature of your world. Our people are hungry and won't survive the winter without your help. I'll make a blood offering when…" Moctu paused and blew out a breath. "…if I get back to shore in your honor and his."

Moctu took it as a good sign when, at that moment, two bonfires lit up on shore.

"Thank you, Sea Spirit," he mouthed silently.

Shortly thereafter, Palo and Nabu caught up to Moctu's boat, and they hitched theirs together, increasing the drag.

The seal began to tire, and Pele gradually pulled the cord in until the rounded form of the beast could be seen on the surface. He quickly tied off the cord, then raised his bow and shot two untethered arrows into the huge animal.

Pele raised one hand toward the Warrior stars, and in his language, said words that were obviously an expression of gratitude. Moctu, as well, took hold of his wolf talisman, bowed his head and offered a prayer of thanks to both the Sea Spirit and to his Wolf Spirit totem.

It was completely dark by the time they had the seal well secured to both boats, and they started paddling toward the two tiny lights on the far-distant shore.

"Need paddle fast," Pele warned. "Other fish, big fish…" Pele put one hand outstretched, sticking up from between his shoulders, and struggled to find a word.

"Sharks?" Moctu suggested. "You think sharks will come?"

Pele pointed excitedly toward him and said, "Sharks. Yes, sharks. They smell blood, they come."

With markedly higher intensity, the four boatmen paddled hard, and Pele began singing a song to synchronize their stroking.

It began, "Yo ho, teng so meh han," then "Wo ho, seng poto pan." Then Pele sang words in between, and repeated the chorus.

Over and over, he sang it. Moctu, Palo and Nabu understood none of it, but the song did serve to coordinate their paddling.

Chapter Ninety-Four

"Help me put more wood on the fires," Nuri said, shaking her hands for attention from the people loitering by the blaze for warmth. A few broke away to help, and Nuri went back to gnawing on the stub of her thumbnail as she scavenged for wood.

Even with the stars and half-moon, no one had seen any evidence of their four boatmen since the sunset when they were being pulled seaward. Darkness blanketed everything, and all four were feared lost. More and more, louder and louder, Elka heard the word 'cursed' in the grumblings of the crowd.

"Are they right?" she murmured to herself, her stomach churning with fear for her father, her mate and her friends. "Did I cause this too?" Elka put her hand to her stomach in a bid to stem the churn, and she felt the baby kick hard.

"I know, little one, I'm nervous too."

Siduri moved to the waterline and screamed to the dark horizon, "Palo, Palo—come back. Palo, I'm bearing your child. I was going to surprise you, but…you have to come back." She sank to her knees and began weeping. Several people rushed to console her, but she angrily brushed their hands away. She stood and screamed again, "Palo, come back. Please come back. Spirits, please, bring him back." Then she fell to the ground and wept hysterically.

Pausing with an armload of wood, Nuri stopped and tried to contain her tears. She threw her the wood onto the fire, then covered her face with her hands.

"Spirits," she said finally. "We've come so far and weathered so many challenges." She shook her head trying to keep from crying, but a tear rolled down her cheek. "Please bring him back. Please bring them all back. Oh Spirits, please."

"Whose idea was this?" a man grumbled. "Going out there in those flimsy boats."

"We're cursed," said another, and there were mumblings of agreement.

"Listen," screamed Leuna from the water's edge where she had been trying to console Siduri. "Listen, people listen."

"What is that?" Alta said straining to hear over the sound of waves rolling in.

"It's…Is it singing?"

Nuri wiped her face and rushed to the shore, moving out waist deep in the cold surf to try to better hear. After a moment, she turned to the others at the shoreline and burst into tears.

"It is," she said between sobs. "It's singing, and there's…there's more than one of them."

Shortly thereafter, they could all hear it—a deep-voiced man singing foreign words, followed by a group singing a rousing chorus, "Yo ho, teng so meh han," followed by "Wo ho, seng poto pan."

"Palo," Siduri cried, braving the cold surf to join Nuri.

And then the boats were there, or nearly so, scraping bottom on an underwater sand bar half an arrow flight from shore. Still singing, the four men piled out into the surf and began to lug the huge seal carcass and their boats towards the beach.

Screaming with delight and relief, Nuri, Elka, Siduri, and others charged out to meet the men and help them bring the boats and their prize to the beach. Straining to pull the carcass to shore, the seamen quit pulling long enough to embrace their squealing, laughing mates and whoop and jump up and down with them in celebration.

Once on shore, Pele expertly cut the seal open and offered chunks of the blubber for people to eat raw. Everyone took part, some not liking the strong, oily, fish flavor, but all recognizing that it was rich in life-sustaining nourishment. Some found that combining the chunks with a little seawater helped make it more edible.

Soon, seal steaks were being roasted, and the group feasted the rest of the night. The beast weighed as much as four or five adult men, so it would provide feasting for several days.

True to his word, Moctu moved away from the fires, and he knelt in the sand just below the debris line of high tide. Cutting the top of his hand with his knife, he trickled blood into the wet sand.

With Nuri looking on, he said, "Thank you, Sea Spirit for your generosity. You've given our people a great gift—the gift of life for us. We use this gift with respect as we care for our friends and families. We honor you especially as this cold winter bears down on us, and we hope you'll help us with further gifts. I offer this small blood token as a sign of our gratitude."

Early the next morning, Pele roused a still-weary Moctu from where he lay next to Nuri in comfortable furs. Pele's breath clouded in the wintry, seaside air as he said, "Need go fish. Best in morning. We um…" His hands made a 'getting up' motion, and he nodded toward where Nabu and Palo slept with their mates. "Palo and Nabu," he said.

Nuri's warm body and the soft, thick furs felt so good that Moctu had to fight off the rebellion inside himself. He rubbed his forehead, trying to wipe away his drowsiness.

"Okay," he said finally. "I'll try to wake Palo and Nabu."

Shortly thereafter, three lethargic men shivered as they watched Pele take the flat board from his boat and cut seal entrails into pieces.

"Chum," Pele said. "Fish like. They come, we catch."

Chapter Ninety-Five

"AND THAT'S HOW IT worked," the old storyteller said to the four young people gathered in the warm hearth. Rubbing the last carved bone on her necklace she added, "That morning, after chumming the water, they caught fish with their first cast of the net. Lots of fish."

"Were they big fish?" the light-haired boy asked.

"Just let her finish the story," his twin sister said, putting a hand toward his mouth to silence him. He looked at her, annoyed and elbowed her hand away. But before he could reply, the old woman cleared her throat and went on.

"Mostly dorado which is about this big." She held her hands a little more than a foot apart. "And it tastes..." She kissed the fingers of one hand and smiled as she flicked the kiss into the air. "And that afternoon they caught more fish—mostly dorado, but also a small tuna."

"And did people see that the curse was broken...that there never was a curse?" the sister asked.

The old woman beamed and nodded. "You've listened well, and you ask good questions," she said to the girl.

The woman studied the alert faces and eyes of each of the young people in turn, then smiled in contentment.

"You've all listened well," she said nodding. Her gray eyes sparkled as she touched the light-haired girl and then her twin brother each on the shoulder.

"And that...was your grandfather," she said with pride. She looked down for a long moment lost in thought. Then nodding, she twisted to pat the old wolf that slept beside her. "That was your grandfather...my Pele. *He* saved our people."

Elka's wrinkled hand trembled as she used a finger to wipe an eye that brimmed with tears.

"What happened to Devu?" her grandson asked.

The boy next to him followed with, "Did the Lion People stop burning the land? Did the burned lands recover?"

"Those are good questions," Elka said smiling. Then she waved them off with a shaky hand. "But we'll save that for next time."

Notes and Annotated References

I would like to thank the following sources for providing information critical to the prehistorical accuracy of *Sleeps With Wolves*:

1. Reich, D. (2018) *Who We Are and How We Got Here*, Pantheon Books. Describes techniques for speeding the processing of DNA and the ramp-up in numbers of genomes analyzed. Examples show how **ancient DNA is dramatically transforming our understanding of human prehistory**.
2. Higham, T. (2021) *The World Before Us*, Yale University Press. An excellent summary of the ideas, discoveries, and technologies employed that are revolutionizing our understanding of the interactions between human species and our evolution. It includes many amazing and eye-opening anecdotes—my favorite being the discovery of **"Denny," the only known, first-generation hybrid of two human species**.
3. Vincent, James (2019) AI helps discover new geoglyph in the Nazca Lines, *The Verge*.
4. Sakai et al. (2024) AI-accelerated Nazca survey nearly doubles the number of known figurative geoglyphs and sheds light on their purpose, *PNAS* 121 (40). First described by Spanish explorers in the 1500's, about 430 Nazca geoglyphs had been found by 2023. Sakai's team trained an AI system to recognize faint traces of geoglyphs, and they identified and field checked 303 previously unknown geoglyphs in a six-month period.Farmer, M. (1994) "The origins of weapons systems," *Current Anthropology* 35, no. 5: 679-681. Projectile weapons used by EMHs in the Levant circa 40,000 years ago were from darts thrown by spear throwers.

5. Andrea, A. editor, (2011) *World History Encyclopedia*. The bow and arrow were probably invented in the Late Upper Paleolithic, perhaps as early as 18,000-20,000 years ago. The best evidence for this comes not from the bows but from projectile points that were too small to place on spears or darts.
6. Blitz, J. (1988) Adoption of the bow in prehistoric North America, *North American Archaeologist*, Vol. 912, 127. There is clear evidence for bow in North America only from 3000 BC.
7. Vutiropulos, N. (1991) Long-range weapons in Southeast Europe—The Neolithic to Early Bronze Age, *Internationale Archäologie* (C. Dobiat, ed.) Research indicates that the mechanics of bow and arrow were not developed before the Neolithic.
8. Metz, L., Lewis, J., Slimak, L., (24 February 2023). Bow-and-arrow, technology of the first modern humans in Europe 54,000 years ago at Mandrin, France. *Science Advances*, 9 (8). **There is good evidence for bow and arrow use at Mandrin, France 54,000 years ago.**
9. Slimak, L.; Zanolli, C.; Higham, T.; et al. (2022). Modern human incursion into Neanderthal territories 54,000 years ago at Mandrin, France, *Science Advances*. 8 (6).
10. Brown, K., et al. (2012) An early and enduring advanced technology originating 71,000 years ago in South Africa, *Nature* 491, 590-593. **Heat-treated microliths found in South Africa dating back 71k years ago, were likely used on arrows shot by bows. This is among the oldest-known cases of likely use of bow-and-arrow technology.**
11. Zimmer, C. (2024). Fossil Trove From 74,000 Years Ago Points to Remarkably Adaptive Humans, *The New York Times*. The **oldest known arrowheads likely date to 74,000 years** ago in Ethiopia.
12. Lombard, M., and Shea, J. (2021) Did Pleistocene Africans use the spearthrower-and-dart? *Evolutionary Anthropology*, Wiley Online Library vol 30, Issue 5. Evidence keeps building for ever-more ancient bow-and-arrow use.

13. NTB: Comparison of references 5–8 with references 9–12 demonstrates that **research is mounting in favor of a much earlier use of bow-and-arrow technology (certainly by 45,000 years ago, at least in scattered cultures).** These earlier dates add to speculations that we outcompeted or even killed off the Neanderthals and other human species we encountered. Noted anthropologist Sally McBrearty agrees, saying that if we had the bow and arrow, we would have been more than a match for any group we met.

14. Shipman, Pat (2017) *The Invaders—How Humans and Their Dogs Drove Neanderthals to Extinction*. Belknap Press. Domestication of wolves sometime before 40,000 years ago may have helped modern humans overcome Neanderthals, as well as other predatory animal species in Europe.

15. Gross, Michael (2015) Are Dogs Just Like Us? *Current Biology* Vol. 25:17. DNA Dates Dog Domestication Back 33,000 Years.

16. Thalmann, O., et al. (2013) Complete Mitochondrial Genomes of Ancient Canids Suggest a European Origin of Domestic Dogs. *Science*, Vol. 342, Issue 6160, pp. 871–874. A comparison of the genomes of ancient dog and wolves with modern canines, concluded that dogs evolved in Europe 19,000 to 32,000 years ago.

17. Botigué, L., et al. (2017) Ancient European dog genomes reveal continuity since the Early Neolithic. *Nature Communications*, Vol. 8. The oldest, unequivocal evidence of a domesticated dog is 14,700 years ago in a German site. Using DNA, this study narrowed the timing of dog domestication to 20,000–40,000 years ago. According to their analysis, dogs diverged from wolves between 36,900 years ago to 41,500. years ago.

18. Harrington, Spencer P.M. (1999) Human Footprints at Chauvet Cave, *Archaeology Magazine* from Archive, 2017. Footprints in Chauvet Cave in France of a youth and his or her wolf-dog run for ~150 feet and date to 26,000 years ago. This is the oldest clear evidence of canine domestication in the world (and another reason to love Chauvet Cave).

19. Rappengluck, M. (2004) A Palaeolithic Planetarium Underground—The Cave of Lascaux (Part One), *Migration & Diffusion*, Vol 5. Makes a strong case that **Pleiades (an easily recognizable group of six or seven stars) is represented on several of Lascaux's bulls**, either within the animal or just above it.

20. Sweatman, M., & Coombs, A. (2019) Decoding Europe's Palaeolithic art: Extremely ancient knowledge of precession of the equinoxes. *Athens Journal of History*, 5(1). Used computer software to determine the position of constellations in the Paleolithic European night sky. Upon comparing their star map with the cave images, they found clear alignment. The researchers suggest that **many of the 600 pictures at Lascaux and much of the artwork in other European caves represent constellations**, and the ancients used the images to record dates based on precession of the equinoxes.

21. NTB: Ancient peoples have used stars extensively for directions and time-telling. For instance, the North Star (Polaris) is always to the north and can be seen in all months of the year. In Europe, the Pleiades star group (noted by various ancient cultures as having six or seven stars) reaches its highest point in the sky about midnight during the month of November, making it visible throughout the night during this time.

22. Weyrich, L., et al. (2017) Neanderthal behaviour, diet, and disease inferred from ancient DNA in dental calculus. *Nature*, Vol. 544, 357–361. Neanderthals discovered penicillin more than 40,000 years before we did. Analysis of the DNA of dental plaque of a Neanderthal from El Sidron, Spain shows that he **used poplar bark (which contains an aspirin-like compound) and fungus containing penicillin to treat an abscessed tooth**.

23. Buckley, S., et al. (2013) Neanderthal Self-Medication in Context. *Antiquity*. 87 (337): 873–878. El Sidron Neanderthals likely also used yarrow and chamomile as medicines.

24. *Italian Connection Website*—Wildflowers in Italy—The Magical Mandrake Plant. The **mandrake plant** has bright purple blossoms and (mandragora autumnalis) blooms from September to December in the south of Italy. It **has bulbous roots that can look like a human body. It's hallucinogenic and narcotic, and it can induce a state of unconsciousness—so, it was used as an anesthetic for surgery in ancient times**.

25. *Stanford Medicine Website*—Garlic. Garlic contains allicin which is a strong antibiotic. Allicin is released when cloves are crushed or chewed. **Garlic has been used as an antiseptic, antibacterial, and antifungal agent**. There is some evidence that it helps the body resist or destroy viruses and other microorganisms by boosting the immune system. It's also been used to fight infection.

26. Drobnik, J., and Stebel, A. (2017) Tangled history of the European uses of Sphagnum moss and sphagnol, *Journal of Ethnopharmacology*, Vol. 209. **Sphagnum moss has been used for centuries as a dressing for wounds**, including throughout World War I. It helps to staunch bleeding and it keeps the pH level low around the wound, inhibiting the growth of bacteria.

27. *Shroomery website*: The Liberty Cap (Psilocybe semilanceata), is the most widespread wild **psilocybin mushroom** of the world. It grows in north temperate areas but it has been found even in Peru, India and also **at altitudes of 4,000 meters in Italy**. Its effects include pupil dilation, spontaneous laughter and delirium. Higher doses cause visual effects and a calming of the body.

28. Guatelli-Steinberg, D., et al. (2016) What Teeth Reveal About Human Evolution. *Cambridge University Press*. Found 39 percent of Neandertals had hypoplasia (grooves in tooth enamel which indicate periods of famine or poor nutrition).

29. Pontoppidan, E. (circa 1752) Attempt at the Natural History of Norway. Vol I and II. The **inner bark of trees**, including white and scotch pines, elm, ash, birch and aspen, is edible and can be ground into flour for use in bread.

30. Elliott, Barkeater, *UrbanDictionary.com*. Barkeater is the English translation of the Mohican word **"Adirondack." Mohawks used it derisively for Algonquian-speaking tribes who ate the inside bark** (cambium) of the white pine when food was scarce.

31. Kobusiak-Prokopowicz, M., et al. (2016) A suicide attempt by intoxication with Taxus baccata leaves and ultra-fast liquid chromatography-electrospray ionization-tandem mass spectrometry, analysis of patient serum and different plant samples: Case report, *BMC Pharmacol Toxicol.* 2016 Aug 30;17(1):41. **Yew is one of the most frequently reported poisonous plants causing fatal outcomes**. 1.75 oz of yew leaves is fatal to an adult human.

32. Roser, Max (2013) Ethnographic and Archaeological Evidence on Violent Deaths, *Our World In Data*. Fascinating compilation and comparison of **violent death** rates in various ancient cultures—documents that many hunter-gatherer societies had high rates (e.g. 1,325-year-old Crow Creek site in North Dakota shows 60 percent and 13,000-year-old site in Nubia indicates 46 percent!).

33. Otterbein, Keith (2004) *How War Began*. Texas A&M University Press. **War was common in prehistoric cultures**.

34. Martin, Debra, and Frayer, David (1997) *Troubled times: violence and warfare in the past*. Gordon and Breach Publishers. **Twenty five percent of men died from warfare during most of human history and prehistory**. Documents evidence of violence in ancient bones, but also uses case studies in archaeology, ethnology, and osteology, to show **prehistoric cultures were often violent**.

35. Pinker, Steven (2011) *The Better Angels of Our Nature: Why Violence Has Declined*. Viking Books. Presents volumes of data that demonstrate that violence has been in decline for many thousands of years. **Hunter-gatherer societies had high rates of violent death**. The change to an agricultural culture brought a reduction in the chronic raiding and feuding and resulted in a fivefold decrease in rates of violent death.

36. Ember, Carol (1978) Myths about hunter-gatherers. *Ethnology*, 17, 439–448. Excluding equestrian hunters and those with 50 percent or more dependence on fishing, warfare is rare for only 12 percent of the remaining hunter-gatherers. **Ancient hunter-gatherers were not peaceful, with most engaging in warfare every two years.**

37. Feeney, J., (2019) Hunter-gatherer land management in the human break from ecological sustainability, *Sage Journals* Impact Factor, Vol. 6, Issue 3. **Many early hunter-gatherer strategies were not ecologically sustainable and caused environmental degradation.**

38. Trotter, M. and McCulloch, B. (1984) *Unearthing New Zealand.* In less than 1000 years on New Zealand, the Maori managed to drive at least thirteen flightless bird species to extinction by overhunting and destroying their forest habitat with fire.

39. Scherjon, F., et al. (2015) Burning the Land: An Ethnographic Study of Off-Site Fire Use by Current and Historically Documented Foragers and Implications for the Interpretation of Past Fire Practices in the Landscape. *Current Anthropology* 56(3). Off-site **fire use for hunting probably dates back to when fire was first acquired**. Its use had both beneficial and detrimental effects.

40. McCall, G. (1994) *Rapanui: Tradition and survival on Easter Island.* Allen & Unwin Publishers. **Polynesians colonized Easter Island** around 400 AD, and in roughly 1,000 years they (and the rats they brought) had **completely deforested the island and caused the extinction of many species.**

41. Whipple, C. (2007) Hunter Gatherers: Not-So-Eco-Friendly, *Swarthmore College Environmental Studies Website.* **Hunter-gatherers** generally worked an area until the resources were used up, then moved on to a new area, where they would live until resources again became low. This way of living was **far from environmentally conscious**.

42. Von Petzinger, G., *The First Signs*, Simon & Schuster. A fascinating book that explores the mysteries and similarities of the symbols left behind by ancient people in caves all over the world. Chapter 13 deals with a necklace found in a German cave along with female remains that dated to 16,000 years ago. **The necklace was made from seventy-one deer teeth that had been polished and engraved with a variety of symbols**. Von Petzinger believes the necklace **may have been a memory device to help a storyteller remember major story elements and the order in which to tell them**.

43. Simonti, C. et al. (2016) The phenotypic legacy of admixture between modern humans and Neanderthals. *Science* 351, 737–741. **Genes we got from Neanderthals give us proclivities for depression** and skin lesions resulting from sun exposure (actinic keratosis).

44. Dannemann, Michael and Kelso, Janet (2017) The Contribution of Neanderthals to Phenotypic Variation in Modern Humans. *Amer Jour Human Genet.* Oct. 5; 101(4): 578–589. **Neanderthal DNA affects skin tone and hair color, height, sleeping patterns, and mood in modern humans.** Sun exposure likely shaped Neanderthal phenotypes, and more than half of the identified alleles are related to skin and hair traits. These Neanderthal alleles contribute to both lighter and darker skin tones and hair color, suggesting that Neanderthals were variable in these traits.

45. Mick, J. (2014) Neanderthal breeding was hard but yielded benefits, *DailyTech*, Science. Although interbreeding with Neanderthals may have given modern humans better immunity and disease resistance, Neanderthals seem to have passed along some harmful genes, as well. Studies suggest that genes associated with **increased risk of lupus, biliary cirrhosis, Crohn's disease, and depression were all inherited from the Neanderthals.**

46. Houldcroft, C., and Underdown, S. (2016) Neanderthal genomics suggests a Pleistocene time frame for the first epidemiologic transition, *American Journal of Biological Anthropology*, Wiley Online Library. Disease transmission between human species was likely problematic for all, and the expansion of pathogens from Africa may have played a role in the extinction of the Neanderthals.

47. Varner, G. (2008) *The History and Use of Amulets, Charms and Talismans*, Lulu Press, Inc. Discusses the use of amulets, talismans, totems, and figurines, etc. as well known in Native American culture to provide luck, protection and good hunting. The items are thought to give a direct link with gods or spirits.

48. Gomez-Jeria, J. (2016) Biology and Philosophy. I. The Paleolithic, *World Journal of Research and Review* 4(1). More than 200 Venus figurines have been found, mostly in Southern Europe, dating back at least to 35,000 years ago. **Most researchers suggest that they served a religious purpose related to fertility and/or the Earth Goddess**, but it's unlikely that we'll ever know for sure.

49. Johnson, R., et al. (2020) Upper Paleolithic Figurines Showing Women with Obesity may Represent Survival Symbols of Climatic Change, *Obesity*, Vol. 29. Found that Venus **figurines are less obese as distance from the glacial front increases**. This suggest that **overnourished women became the ideal symbol of beauty**, able to survive periods of starvation and cold climate.

50. Brue, A., et al. (2022) Resinous deposits in Early Neolithic pottery vessels from the northeast of the Iberian Peninsula, *Journal of Archaeological Science*: Reports, Vol. 47. Studies of resin use by Neolithic populations in Italy found ubiquitous use of conifer resin in 168 pottery vessels. Mentions recent research revealing **Paleolithic use of birch bark tar and bitumen for waterproofing**.

51. Franco Cazzini F. (2018) The history of the upstream oil and gas industry in Italy, *Geological Society, London, Special Publications*, vol 465. **Italy has abundant evidence of hydrocarbon seepages. Early populations took advantage of these phenomena, harvesting oil and bitumen from the surface.**

52. Mark, J. (2012) The Battle of Kadesh & the First Peace Treaty, *World History Encyclopedia*. The Kadesh peace treaty of 1274 BC, which is the **earliest known surviving peace treaty**, cemented ties between Egypt and the Hittites. The text concludes with an oath before a thousand gods of the lands of Egypt and Hatti, "witnessed by the mountains and rivers of the lands of Egypt, the sky, the earth, the great sea, the winds, the clouds." If the treaty was ever violated, the oath-breaker would be cursed by the gods who would destroy his people and land. Conversely, if he maintained his vows, he would be rewarded by the gods, who would prosper his people.

53. Pearce, E., et al. (2013) New insights into differences in brain organization between Neanderthals and anatomically modern humans. Proceedings of the Royal Society. **Neanderthals had bigger eyes and more brain space dedicated to vision, so they probably could see better than modern humans.** They also may have been able to see better in low light.

54. Massetti, et al. (2021) Litter Impacts on Marine Birds: The Mediterranean Northern Gannet as Case Study. *Marine Pollution Bulletin*, Vol. 171, El Sevier. Remains of gannets, gulls and more than fifty other species of coastal birds have been found in paleolithic excavations.

55. Guillaud, et al. (2021) Neanderthal Foraging in Freshwater Ecosystems: A Reappraisal of the Middle Paleolithic Archaeological Fish Record From Continental Western Europe. *Quaternary Science Reviews*, Vol. 252, El Sevier. Describes Neanderthal exploitation of sturgeon and other freshwater fish in Belgium, France and Spain. Also summarizes other studies that mention their exploitation of Mediterranean dolphin, seals, seabirds and mollusks.

56. Morales-Perez et al. (2019) Mediterranean Monk Seal Hunting in the Regional Epipalaeolithic of Southern Iberia. A Study of the Nerja Cave Site (Malaga, Spain). *Quaternary International*, Vol. 515, El Sevier. **Paleolithic Neanderthal and Homo sapiens made consistent use of seals and other marine life** from multiple sites along the Iberian coast. Seals were processed for their flesh, blubber and skin.

57. *Wikipedia*: Mediterranean Monk Seal—**Mature male Monk seals are about eight feet long and weigh an average of 710 pounds**. They are currently considered a threatened species, with fewer than 700 individuals.

58. Walker, R, et al. (2010) Evolutionary history of partible paternity in lowland South America. *Proceedings of the National Academy of Sciences*. 107 (45): 19195–19200. **Many Amazon tribes (up to 70 percent) believe that any men who have intercourse with a woman during her pregnancy are partial fathers of the child**. The theory also shows up in New Guinea, and elsewhere, and is called partible (or shared) paternity. A **common belief is that the offspring will grow to express the characteristics of the men who contributed the most semen**, Therefore, a good mother will have sex with several men so her child will enjoy the best qualities of her partners (and be partly the best hunter, best napper, or best storyteller, etc.).

59. Caesar, Julius (circa 52 BC) *The Gallic War*. Wrote about the Celts in England who had beliefs and followed practices related to shared paternity.

60. *Wikipedia*: Yew is one of the best woods for making bows. The heartwood is always on the inside of the bow and the younger sapwood on the outside. The heartwood is best in compression whereas the sapwood is superior in tension.

61. *WikiHow*: **How to Tell Time Without a Clock**. With extended arms stack the four fingers of one hand on top of the four fingers of the other, and measure from the bottom of the sun to the horizon. Each hand represents one hour of time, so each finger is about fifteen minutes. Therefore, three hands and two fingers, for instance, would represent three and a half hours until sunset.

62. Archery Historian: The Best Wood for Bows. **Yew, maple, hickory and ash are among the best native-European woods for bows**.

63. *r/Bowyer*: Multiple entries—Lists good woods for bows, plus it has these observations: Woods should be dried (no green woods used). If you can make an impression in the wood with your fingernail, it's not ready, or it's not hard enough. Be careful not to damage the growth rings while making the stave. The grain should run the entire length of the bow, not run off the bow. Sandstone can be used to smooth and shape the wood.

64. Baraniuk, C. (2015) The Story of Rhinos and How They Conquered the World, *BBC.com Earth*, Woolly rhinos were around in Europe until 10,000 years ago.

65. Romandini, M., et al. (2020) A late Neanderthal tooth from northeastern Italy. *Journal of Human Evolution*, Vol. 147. A 48,000-year-old child's canine represents the most recent evidence of Neanderthals in Northern Italy.

66. NTB: According to 23andMe, I am >2% Neanderthal, placing me in the upper 25 percent of their millions of customers. I learned that my paternal haplogroup is I-M438 which is found almost exclusively in Europe where it occurs in about 20% of the population. Like the character Moctu, men bearing this haplogroup were among the first EMHs to inhabit Europe some 30,000 to 45,000 years ago.

67. *YouTube*: Neanderthals Had BLUE Eyes And BLONDE Hair Says New DNA Research, Feb. 2025. Makes a good case that **many Neanderthals likely had light hair, skin and eye color**.

68. Reilly, P., et al. (2022) The Contribution of Neanderthal Introgression to Modern Human Traits, *Current Biology*, 32(18). Neanderthals admixed with early modern humans (largely in the 40 to 55 kya timeframe), and more introgressed Neanderthal variants are associated with lighter pigmentation than dark. These new genes likely helped darker, modern humans adapt to the higher latitudes. Two of these genes that were selected for in the newcomer populations are BNC2 and OCA2, both of which skin, hair and iris pigmentation. Specifically, a **Neanderthal allele associated with blue iris pigmentation and blonde and red hair color was selected for by modern human populations**.

69. Hanel, A. and Carlberg, C. (2020) Skin colour and vitamin D: An update, *Experimental Dermatology*, John Wiley & Sons Ltd. "SNP rs12913832 is located on a transcription factor binding site within an enhancer residing in an intron of the HERC2 gene but affecting the expression of the OCA2 gene. The homozygous G/G-derived allele of this SNP is **associated with blue irises in Europeans**." HERC2 and OCA2 genes were introgressed from Neanderthals.

70. Sparacello, V., et al. (2025) Projectile weapon injuries in the Riparo Tagliente burial (Veneto, Italy) provide early evidence of Late Upper Paleolithic intergroup conflict, *Nature Scientific Reports* 15, #14857. Documents evidence of a **young man killed with flint tipped projectiles 17,000 years ago in northern Italy**. According to the researchers, the evidence suggests **conflict between different groups of hunter-gathers** and may represent violence stemming from competition over resources and territory.

71. Plisson, H., et al. (2025). Arrow heads at Obi-Rakhmat (Uzbekistan) 80 ka ago? *PloS One*, 20(8). New evidence suggests early hominins used projectile weapons nearly 80,000 years ago. This would be the **earliest known use of bow and arrow technology**. Human fossils there have a combination of both Neanderthal and Homo sapiens characteristics.

www.ingramcontent.com/pod-product-compliance
Lightning Source LLC
LaVergne TN
LVHW030752030626
840010LV00013B/8/J
* 9 7 8 1 6 4 4 2 8 5 8 3 1 *